W9-DEV-762

EARTHLY REMAINS

EARTHLY REMAINS

The History and Science of Preserved Human Bodies

Andrew T. Chamberlain and
Michael Parker Pearson

OXFORD
UNIVERSITY PRESS

Andrew T. Chamberlain and Michael Parker
Pearson have asserted their moral right to be
identified as the authors of this work.

Published in the United States of America by
Oxford University Press, Inc.
198 Madison Avenue, New York, New York 10016

Oxford is a registered trademark of
Oxford University Press, Inc.

Library of Congress Cataloging-in-Publication
Data available.

ISBN 0-19-521852-3

First published in 2001 by The British Museum Press
A division of The British Museum Company Ltd
46 Bloomsbury Street, London WC1B 3QQ

Designed by Martin Richards

Typeset in Photina
Printed in Great Britain at The Bath Press

Contents

Preface

The idea of the preserved corpse holds a special place within the contemporary psyche. It is something that repels and attracts, fascinates and horrifies. It is problematic and ambiguous, dead but not gone, an arrested state within the natural process whereby our mortal remains should decay, leaving only clean white bones. In the parlance of anthropologists, the preserved body has not completed the rites of passage which the dead must follow from the moment of death until safely on the other side – it is stuck in limbo, neither fully in the world of the living nor entirely in the world of the dead.

Interest in preserved bodies has probably never been greater. Thousands visit The British Museum each year to see Egyptian mummies and the Lindow bog man. Recent discoveries such as the Alpine Ice Man and the 'Ice Maiden' of Pazyryk have attracted enormous media attention. Images of revived corpses are among our most popular myths – as reanimated automata from Frankenstein's monster to the Borg of Star Trek; as vengeful Egyptian mummies released from the tomb; or as zombies raised from the dead and wreaking havoc on the living. Such creatures are always portrayed as adversaries to human heroes; lacking in humanity, they are only partially sentient beings whose life-force is incomplete.

Our world is one in which people are obsessed with bodies, both living and dead. The cult of the good-looking and healthy body is a billion-pound industry; close attention to corpses is paid by doctors, pathologists and other specialists to ascertain causes of death, cures for disease and the overall health of the population. On the one hand there is public concern at the use of bodies and body parts in medical experiments and yet, on the other, there is intense curiosity whenever such material is presented to the public in the guise of art or education. Perhaps our interest in preserved bodies is heightened by the fact that today few people routinely encounter corpses and those who do – such as nurses, doctors, police and undertakers – are something of a closed community. Death has become very distant, not an integral part of life. Public opinion is often vehement and unpredictable, as likely to swing one way as another when faced with the uses made of the human body. What some see as a moral outrage, others accept as an educational and artistic experience.

These contradictory attitudes are encapsulated by reactions to Gunther von Hagens' *Körperwelten* exhibition, which at the time of writing is on display in Berlin (see colour plate 1). Using a technique of plastination – preserving bodily tissues in silicon, epoxy and polymers – von Hagens has treated over 3,200 corpses. He publically displays these anatomical specimens in varying degrees of dissection, as a form of artistic expression. Surprisingly, he has no shortage of bodies donated by their dying owners. As one female donor said: 'The sense of my being must not end with my death. The thought that after my death I could be part of a museum is comforting and fascinates me.' There are certainly many who long to retain a place in the world of the living when faced with personal extinction, and yet the use of these dissected corpses as an art form is a true challenge to our preconceptions of how the dead should be treated.

Views about the display of corpses, recent or ancient, are volatile and varied. Many indigenous communities are not prepared to tolerate the study or exhibition of ancient remains which they regard as belonging to their ancestors. Other people have strong moral, ethical or religious views, either for or against such study. From the archaeologist's point of view, the wishes of the living must be respected and yet the value of ancient preserved remains is not so much in their morbidity, aesthetic appeal or shock factor, but in their often extraordinary capacity for providing information about lost worlds and ancient lives. Unlike skeletons, preserved bodies offer remarkable insights into people's appearance – face, hairstyle, clothing, bodily ornament and tattooing – and their lifestyle, revealed by ingested food and absorbed substances. Such bodies may even reveal the cause of their death and details of the funerary rites by which their bodies were consigned to the earth. An ancient preserved body is a veritable treasure house of information: the archaeologist can apply the skills and techniques of the forensic pathologist to address questions which may rarely be asked of other archaeological evidence.

This book explores the archaeology of preserved bodies in all their shapes and forms, from the remote period of the Palaeolithic up to the present day. We would like to thank present and former colleagues and students at the University of Sheffield for their advice and encouragement, particularly Paul Buckland, Eva Panagiotakopulu, Dawn Hadley, Paul Nicholson and Ian Tyers. We are especially grateful to Teresa Francis for her encouragement to write this book, Laura Brockbank and Karen Godden for their insightful comments and editing, and Natasha Coates for her research on the illustrations. We also thank colleagues at The British Museum for agreeing to read the manuscript in advance of publication: Janet Ambers in the Department of Scientific Research, John Taylor and Renee Friedman in the Department of Egyptian Antiquities, J.D. Hill in the Department of Prehistory and Early Europe, and Nigel Barley, Jonathan King and Colin McEwan in the Department of Ethnography.

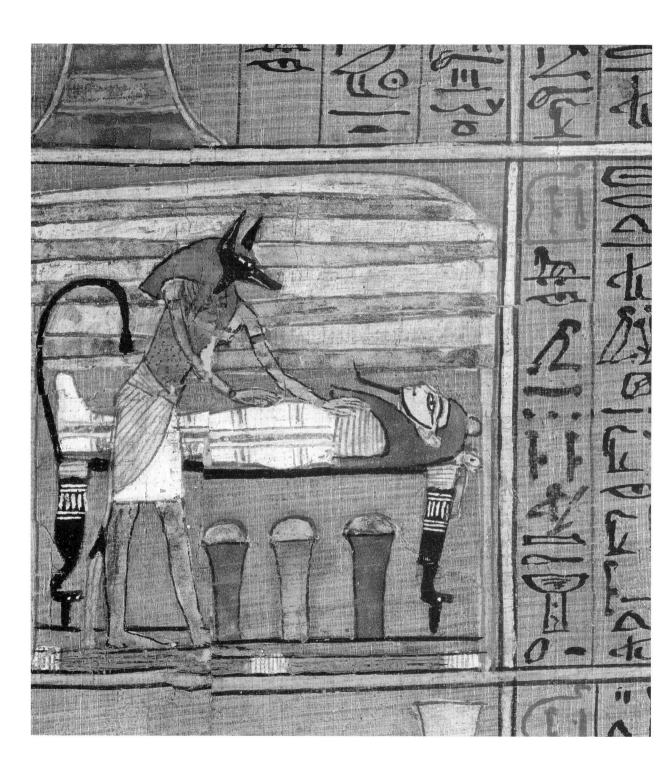

CHAPTER I

Myths and realities of preservation and decay

INTRODUCTION

It is very common to see dead Bodies which have been preserved by Art for many Ages: but that they should lie unchanged by means of the Soil only in which they were Interred, will appear strange to most People.[1]

It has long been known that bodies placed in the ground decay, and that this natural process can be slowed or arrested by human intervention. Nonetheless we are often surprised when bodies spontaneously become mummified, rather than putrefying and decaying in the expected manner. Decay of the body after death is viewed as a normal, natural event, in contrast to preservation which generally requires an explanation. What are the origins of our normative view of what happens to the body after death? Most people live in regions of the world that are predominantly warm and humid for at least part of the year, and these are environments that promote the rapid decay of dead bodies. In contrast, bodies are naturally preserved in what are usually categorized as extreme environments: the frozen Arctic wastes, cold rarified air of the high mountains, hot dry deserts, cold wet peat bogs, and dark underground spaces such as caves, mines and tunnels. The conditions that are conducive to life and agricultural production (sunlight, warmth and humidity) are also favourable for the growth of the micro-organisms that are responsible for the decay and putrefaction of dead bodies. Thus in our preferred habitats we would not normally encounter an ancient body that was accidentally preserved after death.

1. Detail of vignette from the Book of the Dead *of the scribe Ani, showing Anubis tending the mummy within the embalming booth. Nineteenth Dynasty, c. 1250 BC.*

Hunter-gatherers and nomadic peoples are the traditional occupants of extreme environments, and their experiences and beliefs about the body are more likely to provide insights into the phenomenon of preservation after death.

This book is concerned with the natural fate of the body after it dies, and with the diverse ways in which people can and have intervened to prevent what comes naturally. Not only do we consider the mechanics of preservation and decay, but we also investigate the attitudes to the dead which so often motivate the ways in which the living deal with corpses and react to the discovery of ancient preserved bodies. We examine how the practices of deliberate preservation may have originated in people's habitual observations of processes in the natural world, and in their application of storage and preservation technology to prevent decay. We also discuss some of the new scientific techniques that are being used in the study of preserved tissues, and we explore whether traditional values and modern desires can be reconciled in the ethics of displaying and owning the dead.

NATURAL PROCESSES OF DECAY

Then worms will come and eat thee up, on Ilkley Moor baht 'at.
(English traditional song)

For dust thou art, and unto dust thou shalt return.
(The Bible, Genesis 3: 19)

Much of the human body consists of water, dissolved salts, protein, carbohydrates and lipids or oils and fats: only 7% of the body is formed of durable mineralized tissues, the bones and the teeth.[2] During life the body maintains itself: worn-out cells are renewed, the immune system eliminates any harmful invading organisms, and continuous metabolism replenishes any nutrients that have been consumed or lost. At the time of death all of the cells of our body die and an irreversible process of decay commences. Putrefaction is a gradual transformation in which the complex biological constituents of the body's cells are converted into simpler molecules which are mainly liquids, gases and mineral salts. Putrefied soft tissues can literally turn to liquid: much of the body consists of water anyway, and the breakdown products of proteins and fats are either liquid or are highly soluble in the water which percolates through soil. When cells die they release their own stock of enzymes which begin the breakdown of proteins and fats: this self-destroying activity is called autolysis. However, most of the dissolution of the cadaver is undertaken by the enzymes produced by foreign bacteria which rapidly invade all of the soft tissues after death.

There are two kinds of bacteria that are involved in putrefaction: those that inhabit the surfaces of the living body and therefore are present at the time of death, and secondary bacteria that colonize the cadaver from the surrounding burial environment. During life commensal bacteria live on our skin and inhabit our digestive and respiratory tracts. These micro-organisms are not harmful and indeed the bacteria in our intestines are essential for the normal digestion of the food that we eat. During life our immune system prevents these bacteria from damaging the body's cells but after death they begin to invade and consume the muscular, fatty and connective tissues of the body. Other bacteria can migrate into the cadaver depending on the circumstances of burial. Humic soils contain high concentrations of bacteria, and other organisms that feed on decaying flesh, such as insect larvae, may also carry bacteria with them when they visit the cadaver. In some terminal illnesses, a bacterial infection of the blood (septicaemia) can occur, and in these cases the circulatory system disseminates the bacteria to all body tissues immediately prior to death. The cadavers of individuals who die with septicaemia therefore can decay unusually rapidly.

Factors that determine the initial rate of decay of human cadavers are principally those that influence bacterial growth.

The temperature of the depositional environment is an important determinant of decay: most bacteria thrive at an optimal temperature of 37°c, but soil temperatures are typically much lower than this and bacterial growth is greatly inhibited at temperatures below 10°c. The rate of loss of internal heat from the gradually cooling corpse also affects the growth of endogenous bacteria. Large adults lose heat slowly, whereas the bodies of children lose heat more rapidly after death, and as a result children's bodies are often less affected by decay caused by internal bacteria.

'Aerobic' bacteria need both water and oxygen to survive, and burial environments that are dry or that have restricted air circulation tend to delay putrefaction and promote natural mummification. Equally, bodies that have become dehydrated prior to death tend to decay more slowly, as do the bodies of stillborn children who have not consumed any food (their digestive tract will contain only sterile fluids from the mother's womb, a fact that sometimes enables forensic pathologists to distinguish between stillbirths and live-born babies that become victims of infanticide). Burial in a deep grave or in a well-sealed coffin inhibits the process of decay, and bodies that are deposited under

Table 1. Conditions conducive to soft tissue preservation.	
Cold	Chemical reactions proceed slowly. Bacterial activity and the hydrolysis of fats and proteins are reduced. Groundwater circulation may be reduced or eliminated by formation of ice. Adipocere may form.
Dry	Tissues dehydrate rapidly, reducing bacterial activity. Hydrolysis is reduced or eliminated. Salts in dry soils may reduce microbial activity in buried bodies.
Wet	Waterlogged burial enhances preservation of proteins in cartilage, skin, hair and nails. Tanning agents in peat bogs preserve collagen molecules. Reduced oxygen and concentrations of nutrients prevent bacterial growth.
Sterile	Anaerobic conditions and high levels of toxic metals or minerals inhibit the growth of micro-organisms.

water also tend to decompose more slowly. Taken together, these findings imply that preservation can be enhanced in several kinds of environments, including very dry, very wet, very cold and anaerobic (i.e. oxygen deficient) conditions (see Table 1).

Apart from the ubiquitous bacteria there are many other types of organism that are ready to exploit the nutritive potential presented by a human cadaver. Like the aerobic bacteria, fungi require a combination of moisture, warmth and oxygen in order to grow, and under appropriate burial conditions microscopic fungi can rapidly colonize a cadaver. Fungi grow by producing fine filaments called hyphae that develop and spread to form an interwoven mass called a mycelium. The hyphae secrete enzymes that can partially digest organic materials in their vicinity. The breakdown products from this external digestive process are then absorbed by the hyphae and are used to sustain the growth of the mycelium. Fungi can grow on a wide range of organic substrates and they will even dissolve bone and teeth in their search for nutrients.

Animals that specialize in eating decaying matter are called saprophages, and several species of flies as well as some beetles, snails and worms can feed on decaying human flesh. The 'worms' that are historically associated with decaying corpses are in fact maggots, large insect larvae that consume dead tissues before they pupate in order to form a new generation of adult insects. Some carnivorous snails will burrow to considerable depths to seek out decaying organic matter, and their shells are sometimes found inside skulls from Neolithic long barrows. Scavenging mammals such as rats and foxes, and some birds such as crows and ravens, will feed from fresh cadavers, especially if the body is exposed or only partially buried. Bodies deposited in freshwater or in the sea are scavenged by some types of fish and crustaceans (crabs and lobsters).

In all but the coldest environments

2. Maggots of the bluebottle fly Calliphora vomitoria. This species is one of the first to visit a body that is exposed after death.

insects are usually the first animals to discover a decomposing body, and they are the predominant animals responsible for consuming the dead tissues. The cadaver forms a rich but temporary supply of food, and many species of insects can thrive either as necrophages, making their living by eating the decomposing tissues, or as parasites and predators that then feed on the larvae of the necrophages. Forensic scientists have made detailed studies of the rate and timing with which insects invade cadavers, as an aid to understanding events surrounding natural deaths and homicides. Usually the bodies of domestic animals such as pigs are used in these experiments and, in one experiment, more than 400 different insect species were recorded on a group of pig carcasses.[3] Experiments using human cadavers are less common, but forensic anthropologists at the University of Tennessee have established an Anthropology Research Facility (colloquially termed the 'body farm') where donated human corpses are exposed or buried to simulate different conditions of death.[4]

While the soft tissues of the body putrefy and decay, the mineralized parts of the body's skeleton – the bones and the teeth – usually remain intact. However, bones and teeth contain more than just mineral, and both protein (collagen) and fats are major organic constituents of the skeleton. The strong fibrous collagen in bone is efficient in resisting tensile forces, and together with the resistance to compression provided by the mineral (calcium phosphate) these components make bone a very strong but flexible material. Fresh or 'green' bone looks and feels waxy or greasy because of its high fat content, but as the fat gradually oxidizes or hydrolyzes the bone becomes drier in appearance and to the touch. The collagen in bone is more resistant to decay because the protein fibres are surrounded by mineral crystals which protect the collagen from the enzymes released by micro-organisms. However, over long periods of time collagen will react with water in the burial environment, especially when the temperature is warm, and the protein can also be damaged by exposure to ultraviolet light, for example if bone lying on the surface of the ground is exposed to direct sunlight. As the

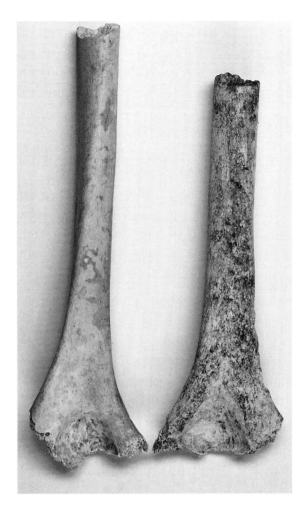

3. *Fragments of two human arm bones showing differing degrees of preservation. The bone on the right has been exposed above the ground for a long period of time and has been partially bleached through exposure to direct sunlight. The better preserved bone on the left is from an excavated burial.*

collagen loses its integrity the bone becomes porous and reduced in strength, and as a result archaeological skeletons are often fragile and 'chalky', with a tendency for the bones to break into irregular fragments. Different bones are more or less likely to decay depending on their size and shape. The large and robust weight-bearing bones of the legs are often well preserved, while the smaller and spongier bones as well as those of irregular shape are more likely to be damaged or destroyed.

Archaeologists have been conducting long-term experiments designed to find out just what happens to organic materials when they are buried under controlled conditions at sites that are constructed to resemble prehistoric earthworks. The experimental earthwork constructed at Overton Down in Wiltshire in 1960 was one of two large artificial earthworks located in southern Britain with the specific aim of finding out how the earthworks, and the materials contained within them, decayed over long periods of time. Sections of the Overton Down earthwork have been partially re-excavated at regular but expanding intervals of time. Excavations were carried out after one, two, four, eight, sixteen and thirty-two years, and the next scheduled excavation is planned to take place in 2024, sixty-four years after the earthwork's initial construction.[5] At the thirty-two-year excavation the investigators found that pieces of bone that had been defleshed and then buried within the chalk fill of the earthwork had suffered from surface cracking, probably as a result of periodic changes in temperature and humidity. There were also some scratches caused by the movement of stones next to the bone, and microscopic study of the samples revealed that the surfaces of the bone were etched and scalloped by the action of saprophytic fungal hyphae: some of these fungal growths were seen during the excavation.

Although the soft tissues of the human body decay progressively after burial, there are occasions when some of the tissues resist decay or become transformed into more stable compounds. Hair is often the best surviving material, as it consists largely of the structural protein keratin which is strong and insoluble. Intact hair is sometimes preserved together with bones and teeth after the rest of the body has disappeared. Adipocere is a waxy or soapy compound, sometimes referred to as 'grave wax', which is formed when body fats are converted by endogenous and bacterial enzymes into fatty acids.[6] The process sometimes occurs under anaerobic conditions but can also take place in warm and dry environments. Adipocere survives particularly well when bodies are deposited in bogs or in graves that are waterlogged, and in bodies that are frozen in ice as well as cadavers sealed in lead coffins. Under these conditions the supply of oxygen may be limited and the deposits of fats in the body are hydrolyzed but do not decompose completely. Adipocere most commonly develops in the subcutaneous fatty tissues of the cheeks, the breasts, the abdomen and the buttocks, but it can also form inside body cavities where it helps to preserve internal organs such as the brain (see below, p. 49).

INCORRUPTIBLE FLESH

Hamlet: *How long will a man lie in the earth ere he rot?*
Gravedigger: *Faith, if he be not rotten before he die (as we have many pocky corpses now-a-days that will scarce hold the laying in) it will last you some eight or nine year. A tanner will last you nine year.*

Hamlet: *Why he more than another?*
Gravedigger: *Why sir, his hide is so tanned with his trade that it*
  will keep out water a great while, and your water is a sore
  decayer of your whoreson dead body!
(William Shakespeare, *Hamlet, Prince of Denmark*, Act V,
Scene I)

Shakespeare's macabre observations on the preservation of
buried corpses are a masterful blend of fact and fantasy: diseased
corpses do have a propensity to decay more rapidly, and water is
essential for bacterial action, but it is unlikely that the condition
of the tanner's skin would have a significant effect on the rate of
decay and putrefaction of his corpse. Dead bodies carry inside
themselves a hidden cargo of microbes, and decay begins from
the inside as well as from the external surface of the skin.
Unfamiliarity with microbial processes has generated many
myths about the preservation of bodies: a contemporary 'urban
legend' suggests that the high concentrations of artificial preser-
vatives and antibiotics in convenience (junk) foods prevent our
bodies from decaying after death. This entertaining and super-
ficially plausible conjecture has no basis in fact, because food
additives are metabolized in the living body and only trace
amounts accumulate in the tissues – these are certainly insufficient to prevent the putrefaction of the body after death
(although the increasingly common practice of embalming may
indeed have this effect).

  There was, however, an earlier and more widespread belief in
medieval Europe that the bodies of Christian saints and other
venerated individuals were incorruptible, this being seen as a
miraculous sign that their holy lives had found favour with God.
In the early Christian traditions, disease and corruption were
associated with sin, and it was perhaps natural to suppose that
the bodies of sinners would decay while those who were spiritu-
ally pure would be preserved after death. Preservation after
death of the bodies of saints occurred even though they had
been buried in the ground without special preparation in simple

4. Catacombs of the Capuchins, Palermo, Sicily. The catacombs were in use from the sixteenth to the nineteenth centuries, and contain the mummified bodies of more than 8000 Capuchin monks and noble laypersons.

wooden coffins, so the accounts of incorruptibility are not simply explained as the natural mummification that is seen, for example, in the bodies of modern Capuchin monks interred in catacombs at Palermo and Comiso in Sicily.[7]

One of the most celebrated instances of incorruption concerns the body of St Cuthbert, an Anglo-Saxon missionary whose life was the focus of many miracles. St Cuthbert was appointed Bishop of Lindisfarne in AD 685, and at his death two years later he was buried in an underground tomb in the church of the Lindisfarne monastic community on Holy Island, off the coast of Northumberland. Eleven years later in AD 698 his tomb was opened and his body was found to be undecayed. St Cuthbert's body was exhumed and transferred to a coffin placed above the floor of the church, and thereafter his remains were moved several times as the Christian community from Lindisfarne relocated to different places in northern England, finally settling at Durham in 995. In 1104 St Cuthbert's body was examined to confirm that it was still incorrupt, and the corpse in its coffin was then placed inside a shrine in the newly built Norman cathedral at Durham. Not only was St Cuthbert's body intact but it had retained the flexibility of its joints, and the body was believed to breathe – a unique and remarkable attribute. A guardian of the saint's body was reported to have trimmed

5. *Fragments of the carved wooden Anglo-Saxon coffin of St Cuthbert, recovered from his tomb in Durham Cathedral in the nineteenth century. The figures inscribed on the coffin represent Christ and the Virgin Mary.*

Cuthbert's hair and nails, and on occasion to have engaged in conversation with him, and the miraculous condition of preservation also extended to the clothing that dressed the body of the saint.

A more prosaic interpretation for incorruptible flesh is that the bodies of celebrated religious figures were more likely to be embalmed or to undergo intramural burial (interment within the structure of a building, rather than outside in a graveyard), and these are circumstances that greatly enhance the likelihood of soft tissue preservation. The closed conditions of intramural burial promote the formation of adipocere (see above, p. 18). The presence of adipocere in the cheeks and other fatty tissues of the body counteracts the usual effects of dehydration and shrinkage of the skin, giving the illusory impression that the dead flesh retains the resilience and smooth texture characteristic of the individual during life.

An additional factor may have been the practice in medieval times of exhuming the bodies of saints from earth graves after a relatively short period of burial. Sainthood was usually established several years or decades after the holy person's death, hence the belated impetus for exhumation and translation of the saint's remains to a more prestigious location. St Etheldreda, founder and abbess of a monastery at Ely in Cambridgeshire, was buried humbly in a wooden coffin in the nun's cemetery at Ely in AD 679 and, when she was exhumed sixteen years later in 695, her uncorrupted body was transferred to a stone sarcophagus within the abbey church. Many other accounts of sanctified incorruption are found in the records of the early Christian church in Britain and Gaul, but these phenomena are mainly confined to northern regions of Europe. In the warm conditions that pertain in the soils of Mediterranean countries soft tissue decay can be completed in just a few years, and it is still the custom in certain rural areas of Greece to exhume bodies between three and seven years after burial and to clean the bones of any adhering remnants of flesh.[8] By contrast, in the colder climates of northwest Europe dead bodies decay less rapidly, and burial in

6. *St Etheldreda, founder and abbess of the early Christian community at Ely in Cambridgeshire. Her body was found to be uncorrupted sixteen years after her death in AD 679.*

the ground soon after death may effectively kill any bacteria that are inhabiting the corpse at the time of death. In the cold waterlogged soils at some of the early ecclesiastical sites in Britain decay would have been only minimal by the time the saints' bodies were exhumed and translated to intramural locations, where a natural process of dehydration and mummification might then take place.

There is a poignant footnote to the story of St Cuthbert: in 1827 his shrine in Durham Cathedral was surreptitiously opened by Dr James Raine, the cathedral librarian, and the Reverend W.N. Darnell, a sub-dean at the cathedral. Their action, which did not have the approval of the cathedral authorities, may have been motivated by a desire to discredit the recently emancipated Catholic church, but it resulted in the discovery that the shrine no longer contained a preserved body. Instead what the investigators found, inside the decorated wooden coffin that had been constructed for Cuthbert's first reburial in AD 698, were the bones of several individuals together with a bishop's regalia and well-preserved embroidered ecclesiastical vestments. Over the intervening centuries the soft tissues of St Cuthbert may finally have decayed, but with the relics of the saint being inextricably mixed with remains of other individuals it is now impossible to provide a definitive answer to the mystery.

MEDIEVAL MUMMIFICATION: THE RISE OF ROYAL FUNERARY RITUALS

And here [in Westminster Abbey] we did see, by particular favour, the body of Queen Katherine of Valois: and I had the upper part of her body in my hands, and I did kiss her mouth, reflecting upon it that I did kiss a Queene, and this was my birthday, thirty-six years old, that I did kiss a Queene.
(Samuel Pepys, 23 February 1669)

7. Effigy of Eleanor of Castile, the queen consort of King Edward I, placed above her tomb in Westminster Abbey, London. Eleanor's body was embalmed before being taken to London for her funeral.

The remarkable preservation for years after death of the bodies of saints revealed to believers the sanctity of these missionaries, martyrs and miracle workers but to today's forensic scientist or archaeologist, it indicates a fortuitous combination of circumstances or deliberate intervention. Many of the saints' bodies may have survived by pure accident but, for those recorded in later periods, there may have been human hands at work. We know that in Britain by the twelfth or thirteenth centuries, the knowledge existed of how to prevent (or at least delay) the natural decay of a cadaver – the process known as embalming. Even if the incorruptibility of the body of a saint was due to divine intervention, human ingenuity could do something along the same lines for the less holy and very important. The preservation

of a saint's body was a final expression and confirmation of spiritual power in life; others who had wielded great power – kings and queens, bishops and nobles – could also avoid the normal process of putrefaction. In the case of a monarch, preservation of the body was attained by artificial means, with a specific motive – to maintain the political and divine authority of kingship.

Curiously, during the eleventh century, the funerals of earlier kings such as William the Conqueror and his son, William II, had been low-key and even shambolic affairs. Far from venerating the royal corpse, the court appeared to treat it as an unpleasant problem: William I's corpse was left unattended for rather too long and then disintegrated when it was being stuffed into its coffin. The burial service for this great king was conducted as quickly as possible because he smelt so bad.[9] Yet during the next century, interest and belief in the political and spiritual properties of dead French and English kings grew enormously and competing religious houses vied for the privilege of burying the royal body in their abbey or cathedral.

Between the twelfth and sixteenth centuries, embalmed royal corpses were displayed during lengthy funeral rites.[10] Embalming was also occasionally practised on the bodies of other members of the elite, especially when the corpse was to travel a long way before burial, such as being brought back from war. Medieval

8. *Harold the Earl of Wessex (riding on the black horse) visits King Edward the Confessor in the Royal Palace of Westminster, shortly before Edward's death in AD 1066. In a later scene, King Edward's body is carried in its funeral bier to the newly built Church of St Peter (later known as Westminster Abbey). Detail from the Bayeux Tapestry.*

embalmers did their best: the deceased's torso was opened up from the neck to the groin, the viscera were removed, and the cavity was then washed with vinegar and packed with salt and spices. By the mid-1300s, the word 'balm' (which in its broad sense refers to any aromatic resinous product) had certainly acquired its specific meaning of 'an aromatic preparation for preserving the dead'. At this period, the corpse was smeared with a paste or ointment containing preserving spices and then wrapped like a mummy in strips of cerecloth – fine waxed linen or silk – whose seams were sealed with beeswax. If the corpse was that of a king, it wore his coronation regalia with reproduction insignia. Dead bishops like Archbishop Godfrey de Ludham (d. 1265), excavated in his lead coffin under York Minster in 1969, wore the clothes and mitre in which they once said mass.

Removal of the internal organs was a technique employed initially as a matter of expediency but, by the thirteenth century, it had become a matter of ceremony and spiritual power. A king's heart and even his entrails brought a special ritual significance to the place where they were buried. When Edward I died in 1307, on his way to hammer the Scots one more time, he was well aware of the power that would reside in his body even after death. He requested on his deathbed that his remains be carried before the English army until Scotland was vanquished. He also wanted his heart to be buried in the Holy Land, but the dead

king's wishes were ignored and the body was brought back to London. It was buried in Westminster Abbey almost four months after his death, with the viscera removed and the whole body preserved and wrapped.

King Edward's tomb was opened in 1774 and a description exists of the condition of the corpse. It had turned a blackish chocolate brown and the face retained its exact form though part of the flesh was 'somewhat wasted'. The desiccated body was shrouded in a coarse linen cloth waxed on its inside, and dressed in a tunic, mantle and stole decorated with filigree and fake jewels. The dead king held two sceptres in his hands and sported a gilded crown on his head. Beneath these trappings the entire body and head, including the individual fingers, were tightly wrapped in a very fine linen cerecloth.[11]

By the time of Edward I's death, royal funerals were lavish affairs in which the corpse was displayed as the focus of ceremonial. Yet this period of public display of the royal body was a fairly brief one. When Edward II died in 1327, on the funeral bier in the place where his corpse might have lain in state, there lay instead an effigy of the dead monarch. This was the first of many such royal effigy burials in both England and France. These life-like effigies of the royal corpse were part of the symbolism of the continuity of monarchy: although the body mortal had ceased to be, the body politic endured.[12]

Embalming continued as a means of preserving royal and upper-class corpses, maintaining the body intact, protected within its close-fitting sealed lead coffin. It certainly worked – the dead queen kissed by Pepys in his diary entry above was over 200 years old! Katherine of Valois, the wife of Henry V, had died in 1437 and her body had been embalmed before burial. Her preserved corpse was dug up during rebuilding work in 1502 and became something of a tourist spectacle. It eventually began to deteriorate, probably due to visitor wear and tear, and was finally reburied in 1878.

Although in life Oliver Cromwell overthrew the monarchy, in death his body was treated (at least initially) like that of a king.

9. *Plaster cast of the death-mask by D. Brucciani & Co. taken from the face of Oliver Cromwell, who died in 1658.*

When Cromwell died in 1658, an effigy was dressed in royal robes with a crown, orb and sceptre although his body had already been buried, some said because his embalming had gone dreadfully wrong.[13] Such displays of dead English rulers, whether actual corpses or symbolic effigies, were not always expressions of esteem and veneration – after Charles II's restoration in 1661, Cromwell's corpse was exhumed from its prestigious grave in the tomb of the kings at Westminster Abbey, dragged through the streets of London and hung from the public gallows at Tyburn. The body was later cut down and buried near the gallows, but his head was hacked off and displayed on a pole above the entrance to Westminster Hall.[14] A skull reputed to be Cromwell's was bequeathed by a private collector to Sidney Sussex College, Cambridge where it was buried near the College Chapel in 1960. The famous wart on the end of his nose supposedly survived, kept at one time in the rooms of the Society of Antiquaries in London though a recent search failed to locate it.

PRESERVED AMERICANS

In the plaza of the temple . . . those charged with this duty would bring into the said square the embalmed bodies of the dead lords. . . . The reason they brought out these dead bodies was so that their descendants could drink with them as if they were alive, and particularly on this occasion so that those who were knighted could ask for the deceased to make them as brave and fortunate as they had been.[15]

Five years before the death of Cromwell, a very different culture made much of its mummified rulers. A Spanish priest, Father

10. Mummified bodies of the Inca king Huayna Capac (d. 1525), his queen, and the body of a servant being carried on a litter from Quito to Cuzco where they were entombed in a funeral sepulchre, or pucullo. The preserved bodies of the Inca elite were tended by custodians and exhibited during religious ceremonies and state occasions.

Bernabé Cobo, chronicled the Inca religion and its customs in the centuries after the conquest of Peru. The scene he describes seems quite bizarre. The embalmed corpses, wrapped in fine clothes, were seated according to seniority around an open square, each waited on by an attendant whose duty was to keep flies off the corpse.[16] The square was full of living people drinking toasts to the dead. Through their attendants, the mummies raised toasts to the living in return.

Cobo also records how the Inca ruler's preserved body was carried in a procession through the fields in order to encourage the coming of the rains. The ancestral mummies were considered to be intermediaries between the living and the gods, giving strength and bravery, and legitimating the power of the living. They were not only used to manipulate power but they were also the very basis on which power rested. An Inca usurper would have had to physically remove the bodies of the ancestors that were the personification of the claims to power of the ruling dynasty.[17] Even ordinary acts such as betrothal required assent from the mummified dead, as Pedro Pizarro (cousin of the conquistador Francisco Pizarro) discovered to his surprise in the century before:

Well, I expected to speak with a living Indian but they took me before the bundle of one of those corpses, where it was placed in a litter, in the manner that they were kept, with the Indian designated to speak for him [the suitor, an Inca captain] on one side and the woman in question on the other, seated next to the mummy. Once we went before the corpse the translator repeated the message, and while we remained somewhat suspenseful and silent, the Indian looked at the woman (which I understand was to understand her will); well, having done as I said the two answered together that her dead lord declared that it should be so, that the woman should be taken to the captain as the apo [Francisco Pizarro] *wished.*

(Pedro Pizarro 1571 [1978])[18]

Just over two centuries after Cobo was writing, the body of another American chief was preserved and paraded for his people to see. Abraham Lincoln, president and martyred hero of his young nation, was embalmed in the President's Room in the White House.[19] The undertaker, Harry Cattell, drained the blood, removed the viscera and employed a chemical that hardened the corpse. European North America had become accustomed to this practice of embalming during the years of the Civil War, which had ended just days before Lincoln's assassination. The corpses of soldiers were often embalmed so that they might be brought long distances from the battlefields on which they died to their homes where their families might bury them.

Lincoln's embalmed body was placed on display in Washington DC, first in the White House and then in the Capitol. On 21 April 1865 it began a twelve-day train journey across the United States, from Washington via New York to his home town of Springfield, Illinois. On this slow journey of 1662 miles the body was taken off the train in each of the many cities, where lengthy stops were made for funeral orations and for viewing by thousands of mourners. The embalming was only intended to be temporary, to ensure that the corpse survived what was effectively a sixteen-day funeral before burial in Oak Ridge Cemetery, Springfield.

Even after burial Lincoln's body continued to attract attention.[20] Moved out of its vault to allow for the construction of a memorial obelisk, the coffin was reopened during this transfer to a temporary vault. Six of Lincoln's former associates were present, to verify the corpse's identity. The corpse was once again viewed and identified in 1871 when it was placed in a new metal

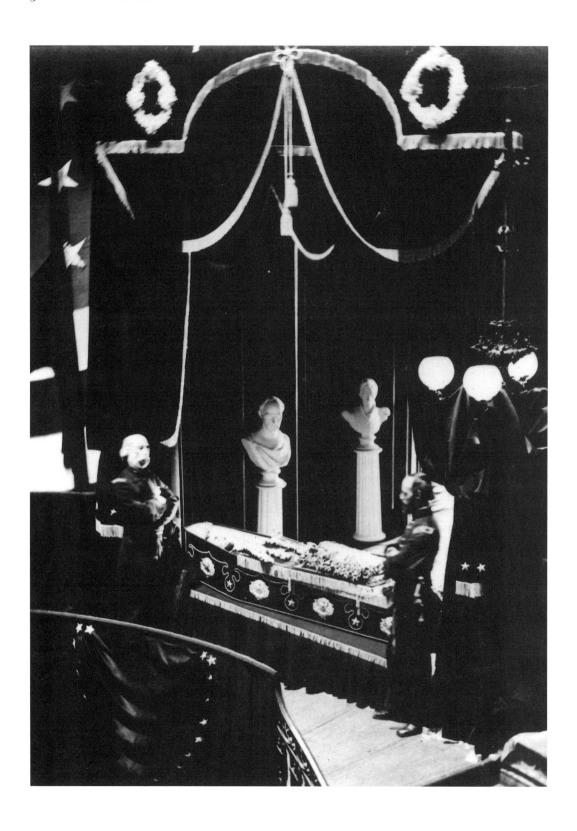

11. *Abraham Lincoln lying in state in City Hall, New York. Lincoln's embalmed body was displayed in Washington and then taken on a twelve-day train journey through America, where it was viewed by millions, before reaching its final resting place at Springfield, Illinois.*

coffin and inserted into the near-completed tomb. In 1876 Springfield police uncovered a plot to steal and hold to ransom Lincoln's body in exchange for an imprisoned counterfeiter. Since then the dead president's preserved body has been protected by the local Lincoln Guard of Honour to prevent it from straying.

COMMUNIST LEADERS AND THE GHOST OF TUTANKHAMUN

He has not died, but merely sleeps:
Our tired leader is resting
Under his granite tombstone
(Adrian Vechernii 1924 *Il'ichu*, 9)

Most people are quite unaware that Abraham Lincoln was the first leader of a modern nation to be mummified by his successors, and yet everybody knows about the body of Vladimir Ilich Lenin – father of the Russian revolution of 1917. After his death in January 1924 his body was embalmed and lay in state in its mausoleum as a place of pilgrimage for millions of Russians. In the months before Lenin's death, after he suffered a stroke, members of the Politburo had begun to argue about what should be done with the corpse. Lenin himself had hoped for a simple burial and some of his successors considered that his death was a good opportunity to break with the past and overthrow tradition, by introducing to Russia the practice of cremation. Trotsky and Bukharin were outraged by Kalinin and Stalin's opinion that disposal should 'conform to the Russian conception of love and veneration of the deceased' and that embalming was a possibility 'in any case for a long enough time to permit our consciousness to get used to the idea that Lenin is no longer among us'.[21] Trotsky and Bukharin could not believe, quite rightly, that Lenin's transformation into a religious relic had anything to do with the secular science of Marxism.

On Lenin's death, his body was first autopsied and then

12. The Russian revolutionary leader Vladimir Ilich Lenin, photographed in Moscow, 23 April 1920.

immediately embalmed as a temporary measure, with the intention of preserving the body until a state funeral and burial could be arranged. A few days later the Politburo decided that the coffin and body would be preserved in a crypt accessible to visitors. The ambient temperature would be controlled at freezing point and the glass-topped coffin kept airtight. The official explanation for this course of action was that hundreds of thousands had queued to see Lenin lying in state in the Hall of Columns at the Trade Union House in the first three days and nights and yet there were many more in Moscow alone who still wished to visit the dead Lenin. It was undoubtedly a popular decision, approved of by the majority of people though opposed by Trotsky, Bukharin and Lenin's family including his widow, Nadezhda Krupskaya. A temporary wooden crypt was constructed outside the Kremlin wall and Lenin lay there for a few weeks until signs of mould began to appear on his skin.[22]

Over the next four months a team of scientists led by the chemist V.P. Vorobiev developed a secret formula for an embalming fluid containing, amongst other things, formalin, glycerine and alcohol. Lenin's body could retain its 'life-like' appearance for decades to come as long as the humidity and temperature within its glass-covered coffin were carefully controlled. In August 1924, seven months after Lenin's death, a wooden mausoleum was opened in Red Square, immediately outside the Kremlin and right at the symbolic and political heart of the Soviet state. In 1930 a massive stone tomb was completed; on the balcony above it, Stalin and later Communist leaders would stand and watch the annual May Day military parades. Beneath it was the laboratory which monitored the coffin's temperature and the body's rate of deterioration, and which was to be used years later for Stalin's mummification.

Lenin's death led to a fierce power struggle in which his corpse was just one of the issues of contention. The legitimacy of the Soviet regime had rested on this single ruler and, with his death, the new Communist state had been in danger of collapsing. Preservation and presentation of the corpse beyond the moment of the funeral was certainly an effective means of consolidating Soviet power as well as the power of Stalin, the victorious contender for the succession. By displaying Lenin's body in this way, Stalin was cementing Communist ideology into Russian nationalism – 'in the pilgrimage to the tomb of Lenin, the rites of Marxism merged imperceptibly into those of the nation-State'.[23] From Lenin's embalmed body, the Bolshevik state was constructing an eternal man-god with a cult-like following.

There were undoubtedly other factors that influenced the decision to embalm. One was religious: the Russian people

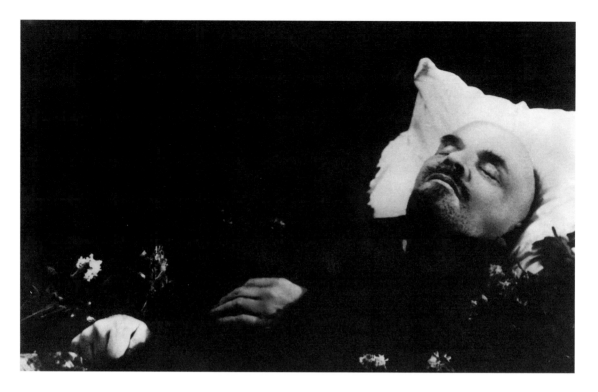

13. *Lenin in his coffin, at the village of Gorky outside Moscow where he died in 1924. Although Lenin and his close friends and family desired a simple private burial, his successor Stalin insisted on the permanent public display of Lenin's embalmed body.*

still held to the Orthodox Church and its popular beliefs about incorruptible remains being a proof of sanctity. Secondly, Leonid Krasin, the man who supervised the embalming and the construction of the wooden mausoleum, was strongly influenced by the writings of the Russian philosopher Fedorov who believed that technology would soon eliminate death, enabling everyone to achieve eternal life and to resurrect the flesh of their ancestors. Finally, the whole world – including the Soviet Union – had been gripped by news of Howard Carter's discovery of the tomb of Tutankhamun.

There is no written evidence of a link between the decision to mummify Lenin and the discovery of Tutankhamun's tomb fifteen months earlier, but there are indications that Tutankhamun was in the minds of those officiating. The Western world was in a frenzy of 'Tutmania'[24] and, despite their ideological isolation, the Soviet leadership was well aware of the discovery – the lengthy removal and recording of the tomb's contents was proceeding as Lenin lay dying. In an interview, one of

Lenin's embalmers compared his colleagues' achievement to that of the ancient Egyptians[25] and, in another press article, Lenin's funeral was compared to those of the founders of ancient states. It seems most likely that the 3000-year-old mummy discovered in Egypt did have some impact on the decision to embalm Lenin. We cannot expect it to have been the prime factor which influenced the Soviet authorities' decision but, together with the other circumstances, it must have provided that happy synchronicity and satisfactory justification – perhaps not even consciously acknowledged by Stalin and others – for taking the path to preservation. In such a way, the ancient past came to inhabit the present and to impress the modern world with links between bodily preservation, god-like power and immortality. Egyptian mummies were no mere antiquarian curiosities but quarries for powerful metaphors about building new states and a new Communist world order in the twentieth century.

Of course, in the post-Communist era one of the most pressing issues for Russian politicians is the question of what to do with 'the smoked fish', Lenin's body.[26] In 1989 the Russian scholar Yuri Karyakin called for the body's removal and 'decent burial'[27] and, after the attempted coup against Gorbachev in August 1991 there was speculation in the press about the possibility of burial.[28] The body is also thought to be in need of further treatment; more work is said to have been carried out in November 2000, but at the time of writing there has still been no final decision as to its fate. Should Lenin's body be buried? Should it be displayed in a less significant location? It is likely that both the condition of the body and also the development of a new political system in Russia mean that action will have to be taken soon.

Whatever happens next to Lenin's embalmed body, the successful preservation of it thus far has exerted a huge influence on the personality cults of other Communist leaders, directly affecting the post-mortem treatments of Stalin, Ho Chi Minh and Mao Zedong. Lenin was merely the first of many twentieth-century public figures whose bodies were embalmed for political

14. *The body of the Russian leader Joseph Stalin, lying in state in Moscow, March 1953. As with his predecessor, Lenin, Stalin's body was embalmed and displayed to the public. The body was then placed in Lenin's mausoleum but later it was removed and buried alongside the graves of earlier Russian revolutionaries outside the walls of the Kremlin.*

purposes. Just as Stalin had overseen Lenin's embalming in 1924 so Khrushchev was put in charge of Stalin's funerary arrangements in 1953. After trepanation of the skull to remove the brain during the autopsy,[29] the embalming of Stalin was carried out by specialists in the laboratory of the Lenin Mausoleum. They dressed the corpse in his marshal's uniform with gold buttons and 'hero's stars'. In a direct imitation of the lying-in-state of Lenin, Stalin too was put on public display in the Trade Union House's Hall of Columns and visited by thousands from all over the Soviet Union. His body was then moved to Lenin's tomb; here in Red Square surges in the huge crowd led to many hundreds being crushed to death – Stalin's final 'blood sacrifice' as it has been called.[30] Stalin was installed in the crypt next to Lenin's body but was later removed and buried nearby, in the area outside the Kremlin wall where the remains of earlier revolutionaries and subsequent Soviet leaders lie.[31]

In 1976 the body of Mao Zedong, hero of the Chinese Communist revolution, was also embalmed for display to the people. Specialist embalmers were brought from Vietnam where they had already performed their arts on the body of Ho Chi Minh, the Communist leader who had fought for independence from the French and led the North Vietnamese until his death in 1969. Mao was expected by many to be cremated and he seems not to have wanted such preservation by embalming.[32] Nevertheless, his body was needed by Hua Guofeng and his other political heirs to keep a grip on China. Some 300,000 mourners paid their respects to their dead leader and a million filled the Square of the Gate of Heavenly Peace for the funeral rally, before Mao's body was finally laid to rest in its mausoleum.

These four Communist leaders acquired god-like status through the cult of personality that surrounded them in life and their continuing preservation after death; three are still receiving visitors.

The Busy Body: Evita's Life After Death

*No man who really had cared for his wife would have used
her cadaver for his own political benefit in the way that Juan
Perón did.*[33]

Probably the most extraordinary exploits of a preserved politi-
cian are those of a woman who became a legend – but not a saint
as she had hoped – after her death. María Eva Duarte de Perón –
Evita – charismatic wife of Juan Domingo Perón the Argentinian
dictator, 'entered immortality' in 1952. Various officials had
suggested the idea of mummification to Evita before her death
and a distinguished Spanish anatomist and pathologist, Pedro
Ara, was brought in even as she lay dying so that he might get to
work on her corpse within minutes of her demise.[34] The
embalmed body was displayed in a glass-topped coffin in the Ministry of Labour, where it was viewed by the faithful – the two-mile
queue lasted for thirteen days – and attended by Perón himself.

Political commentators consider that the lavishness and
pomp of Evita's obsequies had more to do with Perón's manipu-
lation of power than with genuine grief, since the couple's
partnership had become something of a marriage of political
convenience. Evita received a magnificent funeral attended by
2 million people[35] after which the body was taken to the
headquarters of the CGT (General Confederation of Labour) to
be placed in a special chapel. Here, at a cost rumoured to be
around $100,000, Pedro Ara embalmed the body for a second
time. The process took an entire year; after preserving solutions
had saturated the body's circulatory system right through to its
capillaries, certain areas of the cadaver were filled with wax. The
whole corpse was then covered with a layer of hard wax.[36]

The story of Evita's 'mummy' did not end there. Although her
presence as a corpse did continue to provide political support for
her husband, Evita's deathly presence was insufficient to help
Perón stave off General Aramburu's military coup three years
later in 1955. The body, fully protected from decomposition, was

15. *Argentinians pay their last respects to the body of Eva Perón, displayed in a mahogany coffin at the Ministry of Labour in Buenos Aires on 2 August 1952. In the foreground, middle, with his back to the camera, the President of Argentina, Juan Perón, stands with members of the cabinet.*

found by the army in the CGT building, not in its chapel or even in its coffin but on a makeshift altar in an office. The new military regime was keen to have the indestructible body 'excluded from the political scene' so that it would not become a focus for opposition. After years of rumours of its clandestine burial or disposal at sea, the body was secretly shipped to Italy in 1957, accompanied by an Argentinian woman who claimed that she was escorting the body of one Maria Maggi, an Italian widow who had died in Argentina.[37] Evita's body was buried under this name in the Musocco cemetery in Milan, in a coffin said to be of silver.

In 1971 an Argentinian colonel, posing as the brother of the deceased widow, arranged for the body to be exhumed. He accompanied the coffin to its next destination – Spain. Perón was living in exile near Madrid. After many years he had finally managed to obtain permission from Franco, the Spanish dictator, for Evita's body to be brought into the country, although by this time he had remarried. When the casket was opened the corpse was alleged to have been found damaged by hammer blows and knife wounds, and covered with quicklime. In 1973 the ageing Perón was invited to become president of Argentina for a second time but he lived for only another nine months.

In Argentina during the years of Perón's exile, Evita had become the focus of a cult, fostered by the corpse's mysterious disappearance. Slogans such as 'Eva lives' were accompanied by appropriation of her name for support of political factions. On Perón's death, his second wife Isabelita inherited a country in the grip of inflation and political unrest – now was the time to bring Evita back into politics. Four years earlier General Aramburu had been kidnapped and executed by an extreme left terrorist group, the Montoneros. This group then kidnapped the general for a second time – they stole his corpse and vowed to keep it until Evita returned to Argentina. Isabelita arranged for the body to be shipped back from Spain.

Evita's return remained secret until her repaired body was ready to lie on display beside Perón's in the chapel of the presidential residency. Evita's corpse could not, however, save Isabelita and she was deposed in another military coup by that junta of generals who terrorized the population with political 'disappearances' and eventually went to war with the United Kingdom over the Falkland Islands/Islas Malvinas. Perón has since been interred in his family tomb and Evita is buried in Buenos Aires' exclusive Recoleta cemetery.[38] Some 5 m deep, below a stone mausoleum, she lies in her glass-topped coffin in a locked steel vault whose alarm is connected to police headquarters. Buried like a pharaoh of old, the corpse is intended to remain forever beyond the reach of future politicians.

The political messages behind each of these nineteenth- and twentieth-century events were not particularly subtle. Great individuals who had transformed their country were accorded

16. *The burial place of Eva Perón in the Duarte family mausoleum in Recoleta cemetery, Buenos Aires, Argentina.*

permanent or semi-permanent preservation after death; this preservation of the body not only made the deceased into an object of veneration but also served to emphasize the permanence of the political regime. The bodies of Lenin, Lincoln and Mao became metaphors (or rather metonyms) for the emergent states and new orders which they had ushered in.

Do these dead kings and politicians have any relevance for understanding the circumstances in which ancient pharaohs and others were preserved? We think so – they may provide useful analogies for the complex political goings-on that accompanied the deaths of ancient rulers whose bodies were preserved long ago. They force us to question conventional understandings of artificial mummification in ancient societies as merely a ritual practice, unquestioned and timeless, enshrined in myth and religious belief. Archaeologists have tended to treat embalming as a self-evident religious institution rather than as a series of innovations in the performance of funeral rites. Those rites and practices of preservation were – once installed – still subject to modification and change. Yet how did they come to be acceptable in the first place? Edward I's body could have been treated in several alternative ways but embalming was the method selected, in large part due to the political circumstances of the moment, which led to decisions to increase the pomp and duration of the funerary ceremonies during which the body should be displayed. Lincoln, Lenin and Mao were all required to 'perform' long after their deaths, to continue to represent the new political regimes which they had initiated whilst alive. Their corpses embodied the political and moral authority of regimes as their successors struggled to maintain power in the face of the physical death of the founder.

Mummification is not simply a manifestation of religious institutions and belief in an afterlife. It can also be a performance for immediate and longer-term political and social effect. Lenin's body, for example, went through three successive embalmings in the months after his death as politicians realized that it would be needed for a longer period of time than initially foreseen, as the political landscape changed around them. In ancient Egypt the practices of preserving pharaohs were later copied by the nobility and eventually, after many centuries, by the common people. A similar mass adoption of the practice of embalming appears to have happened in the United States within just a century of Lincoln's death but, in medieval England, it never extended beyond the nobility and the clergy. Decisions about treatment of the corpse are made at many levels – political, social, economic, religious – even after a particular rite has long become accepted.

CHAPTER 2

Bog bodies

DELIBERATE PRESERVATION AND ACCIDENTAL PRESERVATION

*I have seen a thousand graves opened, and always perceived
that whatever was gone, the teeth and the hair remained of
those who had died with them. Is this not odd? They go the
very first things in youth and yet last the longest in the dust.*
(Lord Byron, 1820, letter to the publisher John Murray)

Archaeologists often draw a sharp line between deliberate
or artificial preservation and accidental or natural preservation. The former requires positive evidence of bodily transformation by direct human interference. For example, the opening
up of the abdomen to remove the internal organs or the smoking
of the corpse might be aspects of artificial preservation as
opposed to simply leaving a body in dry sand to desiccate
naturally. There is no doubt that the embalmed politicians of the
previous chapter are examples of deliberate treatment but, in
many ancient cases, it is difficult for the archaeologist to decide
with absolute certainty whether the preservation was deliberate
or not. This is especially the case for the many preserved bodies
that have been discovered at wetland sites in northwest Europe.

Archaeologists tend to approach preserved bodies of ancient
cultures with a somewhat minimalist attitude, often starting
from the assumption that people may not have understood much
about the world around them and its natural properties. This
caution in interpretation posits that until we can demonstrate
any society's active involvement in complicated techniques of
mummification, we cannot assume that people understood the
process of preservation of dead bodies. Yet perhaps it is

too cautious to deny people's knowledge of their immediate surroundings. We thus recognize a third form of treatment – intentional natural preservation, caused by human exploitation or deliberate enhancement of natural properties and processes.[1] The people of ancient Peru or Predynastic Egypt, for example, probably knew the drying effects of the soil on flesh and skin, from seeing robbed or exhumed bodies or even the accidentally revealed bodies of animals that had died in the desert. Finally, those natural processes that cause the preservation after death of the human body may well have been understood even though people showed no intentions of artificially encouraging or enhancing natural mummification.

It is not just in hot desert conditions that people would have understood that bodies could be preserved for a long time. Cold deserts and frozen environments would also have been regions where people knew about the caching and long-term storage of meat or observed the naturally mummified bodies of long-dead animals. Peat-cutters digging in the bogs of northern Europe, to find peat for fuel, would certainly have noticed the organic remains of wood and other materials that survive deep in the peat in fresh condition. Of course, just how people interpreted

17. *Animal victim of a drought in Ovamboland, Namibia. The bodies of animals that die in desert environments sometimes become dessicated before the soft tissues begin to putrefy. In exceptional circumstances such bodies can be preserved through natural mummification.*

18. *A peat bog at Thorne Moor, South Yorkshire, England. Peat is formed from the partially decomposed and compacted remains of plants that have accumulated under waterlogged conditions. It contains chemicals that are highly effective in reducing microbial growth, with the result that bodies that are buried in peat can be exceptionally well preserved.*

these preserving properties is something that will have varied from culture to culture and which may be difficult for us to find out. Yet we have to be prepared to give past people credit for knowing the likely physical implications of placing human corpses in particular environments. In this chapter, we look at bodies preserved in bogs, and examine the evidence for how they came to be there.

ARCHAIC BRAINS IN WATERY PLACES

The site of Windover on the Atlantic coast of Florida in the USA is one of the earliest cemeteries in North America. Today the site is a shallow seasonal lake that overlies a deposit of peat that has accumulated to a depth of several metres since it started to form at the end of the last ice age about 11,000 years ago. From 7000 to 6000 BC the lake was used by hunter-gatherers as a resting place for more than 170 of their dead. Adults and children had been buried in the waterlogged peat lying on their sides with their legs in a flexed position, and were accompanied by a range of grave goods including artefacts of bone, wood, shell and stone as well as textiles made from woven plant fibres. Although the bodies had not undergone any special preparation to enhance their likelihood of survival, the chemistry of the lake resulted in exceptional preservation of organic materials, probably because of a combination of anaerobic conditions and a high content of sulphur and other minerals in the lake water. The bodies had become skeletonized (had lost their flesh) but in many cases there were recognizable remnants of soft tissues inside the crania including intact brain tissue.[2] The brains had shrunk to about a quarter of their original size but they showed the usual pattern of ridges (gyri) and fissures (sulci) that are characteristic of the

19. Preserved brain inside a skull dating to the nineteenth century. The skull has been cut open using an electric saw, exposing the intact but shrunken brain inside.

surface of the brain, and internally the distinction between the grey and white matter of the brain tissues could still be discerned. Microscopic sections of the brain tissue were made and it was possible to stain these sections with chemicals that react with the cells and fibres of the nervous system, proving that parts of the original brain cells were still intact.

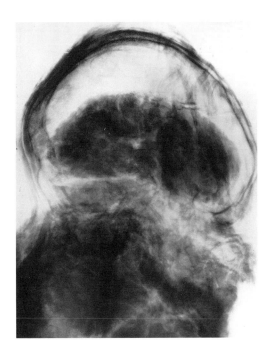

20. An X-ray of the skull of Grauballe Man, showing the brain which has become compact and much reduced in size, a typical finding in preserved bodies.

In most burial circumstances the brain decomposes rapidly. Because of this tendency to decay the brain was often one of the internal organs removed soon after death as part of the preparation of the body for artificial mummification (see Chapter 3, p. 105). Yet, under water-logged conditions, the brain often survives particularly well: preserved brains have also been reported from prehistoric burial sites in flooded caves in Florida[3] as well as in deeply buried bodies in waterlogged soils in some countries of northern Europe.[4] The brain is contained within protective membranes and is also enclosed by the skull, so it has more physical protection than the other organs in the body. But a more important factor is that brain tissue contains high concentrations of lipids and cholesterol and under favourable conditions these can be converted to hydroxy fatty acids which are the major constituent of adipocere (see above, p. 18). The brain is also relatively isolated from the bacteria that inhabit other parts of the body, so although it is susceptible to autolysis caused by enzymes from within its cells the brain does not always putrefy as quickly as the abdominal organs.

Although the Windover bodies had been skeletonized, other bodies deposited in wet conditions can decay in quite a different manner. In the acid environment of peat bogs, the mineral in the skeleton can be dissolved away while the hair, skin, ligaments and some of the internal organs remain exceptionally well preserved. Intact brain tissues have been identified in several bog bodies from Denmark and Germany, and in the bodies found at Lindow Moss in England and Meenybradden in Ireland – in most cases the brain tissue had shrunk inside the skull but, as with the Windover finds, the brain structure was often still visible.

The Bog Bodies of Northern Europe

Peat is an organic deposit that forms when the partially decomposed and compacted remains of plants accumulate under waterlogged conditions. In poorly drained low-lying regions, peat can amass to great depths to form fens and lowland raised mires, while in cold upland areas with high levels of rainfall, blanket bogs develop and can cover extensive areas of ground. In northern Europe peat often contains the remains of *Sphagnum* mosses and it is the special preservative qualities of *Sphagnum* peat that has resulted in many bog bodies being found in Denmark, Germany, the Netherlands and the British Isles. During the formation of *Sphagnum* peat some of the constituents of the plants are converted first into sphagnan (a soluble pectic acid) and then to humic acids.[5] Together these chemicals are highly effective in reducing microbial growth and they also have the remarkable property of tanning collagen fibres, leading to the preservation of tissues such as skin, cartilage, tendon and fingernails. In addition, the generally cold conditions within saturated peat, together with the reduced availability of oxygen within the groundwater and the effective exclusion of carrion-consuming animals enhance the likelihood that bodies buried in peat will avoid decay and destruction.

A hundred or so out of nearly 2000 known bog bodies from northern Europe have been dated by the radiocarbon method and, although all periods of prehistory and early history are represented, nearly half of these dated bog bodies fall in a narrow period of time, between 1000 BC and AD 250. It is to this period (the Late Bronze Age, Iron Age and Roman period) that such 'classic' bog bodies as Tollund Man and Lindow Man belong. We have to take care in interpreting the dating evidence, however. Historical accounts of the discoveries of bog bodies show that they were usually found during peat-cutting for fuel. Throughout Europe the rate of deposition of peat accelerated after a deterioration in the climate that occurred during the first millennium BC, and it was these rich 2000-year-old peat deposits

21. *Tollund Man, a Danish bog body discovered by peat diggers in 1950. The body had been laid carefully on its side in a grave cut into the peat. He had been strangled with a leather cord, which was still in position around his neck.*

that were intensively exploited as a source of fuel during the nineteenth and twentieth centuries. Older peats dating to the Neolithic period or earlier tend to be buried at a depth below that normally reached by manual peat-cutting. Since more Iron Age peat has been dug out, so more finds have been made of Iron Age material. Peat is now rarely dug for fuel but commercial exploitation of the peat deposits has continued to the present day because peat is valued as a sterile medium for use in horticulture, although the belated recognition of the ecological importance of peat bogs has resulted in legislative measures to protect and conserve the few remaining areas of lowland peat.

Although the majority of dated bog bodies are clustered in the Iron Age and Roman period, some people have been discovered in bogs from the Early Neolithic to the age of the Vikings – throughout the period from 3500 BC to AD 1000. Many of them have survived only as skeletons rather than as fleshed corpses but they still provide extraordinary insights into prehistoric society, health and ritual. The earliest 'bog body' known so far is the disarticulated skeleton of a twenty- to twenty-five-year-old woman from the Early Mesolithic, 10,000 years ago, which was found in a peat bog at Koelbjerg on the island of Funen in Denmark. However, bog bodies were uncommon before the Early Neolithic, between 4000 and 3200 BC: from this period there is one body from Stoneyisland in Ireland, another from Hartlepool in England, and a staggering thirty-five bog skeletons from Denmark (see below, p. 54).

Bodies were also deposited in bogs during the Bronze Age, between 2500 and 1000 BC: twelve of these are from the Cambridgeshire Fens and another four are from Holland and Denmark. For the Late Bronze Age to Early Iron Age period there are just two bog bodies from the British Isles: an adult male clothed only in a leather cape in a peat bog at Gallagh, Co. Galway, and the skeleton of a woman with an infant skull in a cut grave from Derrymaquirk, Co. Roscommon in Ireland. As already indicated, many bog bodies from Denmark, Holland, Germany and Ireland date to the Iron Age, from about 800 BC to the first century AD.[6]

Bodies are also found from the Roman period until about AD 500, especially in Germany and Britain. Recent research into over 140 British bog bodies has identified a small group from the Roman period of which four come from Cheshire – the Lindow bodies and another known as Worsley Man.[7] The latter is just the skull of a young man which had a fracture on its top and a garrotte around the neck; the head was severed from its body in

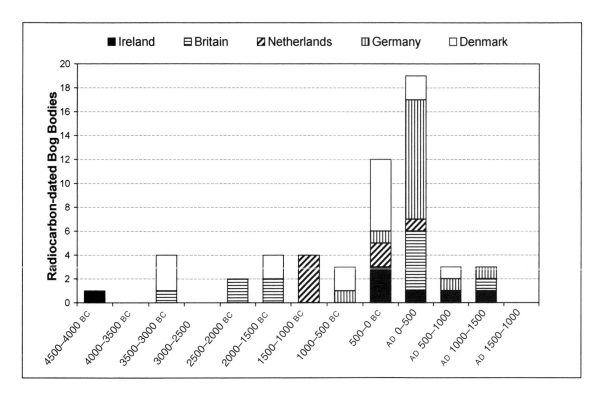

Table 2. *Radiocarbon dates of European bog bodies. The dates range from the Early Neolithic through to the medieval era, but more than half of the bog bodies that have been radiocarbon dated are between 2500 and 1500 years old.*

antiquity which could mean death by decapitation.[8] Just a handful of examples are dated to the 500 years after the Roman period, and a few bodies retrieved from bogs have been firmly identified as belonging to the medieval period. It is possible that the chronological gaps, in the Late Neolithic, in the early first millennium BC, and after AD 500, are significant; we cannot say whether bodies were ending up in bogs throughout prehistory but the dating evidence does so far suggest that there were particular periods when such deposition was especially common.

So how did they get there? Most specialists now consider that the vast majority were deliberately sacrificed or executed. Yet there are other more prosaic interpretations. Bogs and mires are sometimes vast and empty places where people can easily become lost. Did some of them drown? What about clandestine murders? A substantial number of the bodies have injuries, often multiple and severe, which could have been inflicted by killers who lay in wait for their unwary victims strolling through the bogs, and then disposed of the bodies where they would not be found. Other bodies have been found with nooses still around their necks – one unusually imaginative archaeologist has suggested that these were ropes thrown by rescuers, intended to help those drowning in a bog but unfortunately lassoing them fatally around the neck instead. Alternatively, they were perhaps dead already and the noose was placed around the neck to drag them out, but without success. To find out why these types of explanation are not satisfactory, we must look in more detail at the context of the discoveries.

STRANGE BODIES IN DANISH BOGS: A 5000-YEAR-OLD PUZZLE

The Early Neolithic bog skeletons from Denmark have been described as a 'strange group of people'.[9] Although adults and children are present, over a third of these bog skeletons are juveniles, aged between sixteen and twenty. This high proportion of young people is unusual when compared with proportions from Neolithic graves of the period. Some of them have been subject to violence. For example, the skull of a man from Porsmose still has the long bone point of an arrow sticking down through his nose into his jaw; another arrow through his sternum was probably the fatal wound. Another young man from Andenmosen had an injury on his frontal bone, above his left eye, but it could have been inflicted after his death or before. In some cases the bones are scattered, interpreted as evidence of dismemberment and even cannibalism,[10] though post-depositional water

22. *Pathology in the humerus (the bone of the upper arm) in an Early Neolithic bog skeleton from Døjringe, Denmark. The pathological bone is severely shortened compared to the normal bone from the opposite side.*

movement or even post-mortem excarnation (removal of the flesh from the bones) are most likely to have been responsible.

At several of the Danish sites, individuals were found in pairs. One of these pairs from Bolkilde consisted of a sixteen-year-old youth and an adult male. There is no sign of foul play on the younger skeleton but the older man was found with a rope near his neck, suggesting that he may have been strangled.[11] He was also severely disabled; his left leg was stuck at an angle of almost 90 degrees to his body and he would not have been able to move it more than a few millimetres since unusual bone growth deformed his left hip. This deformity may have been the result of some bone-crunching injury that had fractured the neck of the man's femur. He had survived but his hip and leg had been terribly damaged by some tremendous force.[12] Such severe trauma can be caused in car accidents but in what circumstances might it have happened before the invention of the wheel? One possibility was that this man had perhaps been crushed during the moving of heavy objects such as the large timbers or stones erected at tombs and other funerary monuments. More recently, Pia Bennike has found parallels from the last century: this kind of bone growth can be caused by tuberculosis of the joints.[13] She now favours this 'illness' interpretation over the more dramatic one of megalithic injury.

Other individuals in this 'strange group of people' were also disabled, including two men from Døjringe. Each had one arm shorter than the other – in one case a congenital defect and in the other probably due to a lesion which had restricted growth. There was no evidence of the cause of their deaths but both had successfully undergone surgery involving the removal of small portions of the skull. This practice, known as trepanation or trephination, is recorded from the Mesolithic[14] up until the present day.[15] Archaeologists formerly thought that trepanation might be evidence of a belief in magic, being performed by 'primitive' people to release demons

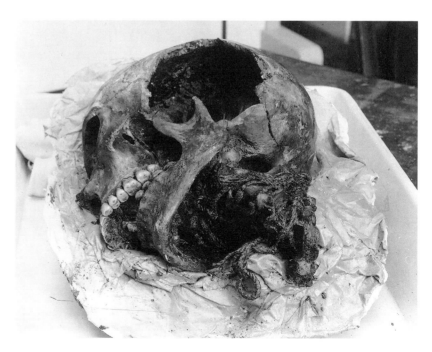

or bad spirits. But the possible reasons for this practice may have been rather different and far more akin to surgery than magic. One of the Døjringe trepanations is oblong and looks as though it was carried out to treat an injury from a heavy implement such as an axe. In neither case at Døjringe did the trepanation cause death – the wounds in the men's skulls have healed edges. Secondly, most trepanations are found on the left side of the skull, the side most vulnerable to right-handed blows in a frontal attack and also the spot most accessible to a right-handed practitioner facing the patient.

These people did not die by accident, alone in the bogs, since a number of the skeletons were accompanied by various items which seem to have been deposited with them. The remains of two individuals were found at Sigersdal; they were accompanied

23. *Trepanation (removal of a piece of bone) on an Early Neolithic skull from Døjringe, Denmark. The smooth margins of the trepanation indicate that the bone healed after the operation was performed. A second area of trepanation is visible at the top left.*

24. (Right) *Skull of the Sigersdal bog body.*

25. (Opposite page) *Deposit of Early Neolithic amber beads found in a peat bog at Sortekær Mose in Denmark.*

by a large pot and one appears to have been strangled by a thick string around the neck. The skeleton of a swan lay with the young woman from Østrup and at Myrebjerg the remains of five people were found with bones of domestic animals. Human bodies appear to have been just one of a range of items – pots, axes, animal bones, complete cattle skeletons, and bone and amber beads – which were deliberately and extensively placed in bogs and open water in the Danish Neolithic.[16] A recent radiocarbon dating programme has shown that many bogs that contain Neolithic pottery also contain Neolithic skeletons.[17] The evidence for shooting or strangulation of three victims, the preferential selection of teenagers, and the interpretation of associated objects as sacrificial or votive offerings together add up to the conclusion that these were socially sanctioned killings, either as fertility offerings or as executions of those who had transgressed social laws and conventions.

IRON AGE BOG BODIES

There are many different ways by which bodies might have ended up in bogs. Clandestine murder, human sacrifice, criminal execution, inter-group fighting, accidental drowning, and post-mortem selection for special burial are all possible considerations. Like their Neolithic predecessors, the Iron Age bog bodies from Denmark are of men, women and juveniles though there are very few children. Not all the bodies provide evidence of a violent end; each tells a different tale about the manner of their death, for which some of the evidence was extracted in the earliest archaeological forensic post-mortems.

In 1835 the body of a woman was found near Haraldskjaer. The corpse was pinned down in the bog with crooked willow sticks and a heavy stake, hammered into the peat. Her clothing included a skin cape, a net bonnet and a finely woven item of sheepswool. Her small hands and feet indicated that she was a person of distinction who had not been subjected to hard work.

A woman from Avningmose was associated with similar crooked sticks and had presumably been pegged down as well. At Huldremose a woman was found, lying on her back with her legs drawn up to her torso. Her head had been shaved and her right arm had been hacked off (almost certainly at the time of her death rather than by a careless peat-cutter) but her left was bound to her body by a leather strap. She wore two lambskin capes and a skirt was fastened around her body by a leather strap and a scarf. An accompanying bladder contained another leather strap and a woollen hairband. A fine horn comb and two amber beads provided useful dating evidence as well as status associations (amber beads in the Early Roman Iron Age of Jutland were linked with women of some standing).

In 1913 the body of a man was found in a bog at Vester Thorsted and sent for forensic examination. Since systematic research into ancient German and Danish bog bodies had already been taking place over the previous forty years, the local judge realized that this might be an archaeological matter and asked the National Museum to take part in the study. Dressed only in a leather coat, the man had been found lying beneath a large branch but there were no marks of violence on the body. The coat was probably of oxhide and was made from many pieces of leather sewn together with leather thongs – comparable to clothing found on other bog bodies in Jutland. The National Museum issued a statement identifying this simple clothing as that of people who were 'not members of respectable society', concluding that this man and others like him must be Gypsies of the sixteenth and seventeenth centuries.[18] Only much later did radiocarbon dates indicate that the man in the patchwork coat had lived some time in the last few centuries BC. By this time, however, P.V. Glob's remarkable book *The Bog People*[19] had become a bestseller and the prehistoric context for this body and others was beyond question (see below).

Several finds of bog bodies were made during the Second World War. A young woman discovered in Bredmose in 1942 had been brought to the bog trussed up and naked, with her legs and arms tightly flexed. She was laid on one cloth and was covered by another. Her hair was plaited into two pigtails, coiled on top of her head and covered by a small woollen net bonnet. A man from Kragelund wore his hair in a knot on the left side at the back of his head; he was wrapped in a sheepskin with cowhide inserts and wore a pair of leather boots. A leather strap and buckle were probably used to hang or strangle him. At Søgårds Mose a man was discovered in 1944 lying face down on

26. Professor Glob preparing the 'tanning' vat in which the Grauballe bog body was treated in order to preserve it for museum display. The body was immersed for eighteen months in a solution containing extracts of oak bark (a natural tanning agent) and then impregnated with glycerin, lanolin and oils to maintain the condition of the soft tissues.

a bed of white bog cotton, wrapped in three skin capes and lying next to a cap and leather shoes. A man's head – previously thought to be a woman – from Roum in Jutland was severed in antiquity and deposited wrapped in a sheepskin.

Glob's book *The Bog People* followed in the wake of a series of dramatic post-war discoveries. In the late 1940s three bodies were found in the bog at Borremose in Jutland. First to be uncovered was a man who had been strangled. The 94-cm-long rope, made of three strands of plaited hemp, was still around his neck

27. *A bog body discovered at Borremose, Denmark, in 1946. A rope made of strands of plaited hemp was found around the neck, and the back of the skull and the right thigh bone had been broken at around the time of death.*

and secured by a slip knot. He had also been hit on the head with such force that the back of his skull was smashed. The second body was that of a woman accompanied by a newborn baby. Her costume consisted of a fringed shawl and a skirt which had been formerly worn as a man's cape (on the basis of pinholes and wear marks).[20] Near her were found a leather thong, an amber bead and a bronze disc. There was half of a pot lying on her right arm. Not only can the pot be dated to the first to second centuries AD but its black burnished surface indicates that it was of a type used by the local elite, an observation supported by the fact that she is wearing an amber bead, an ornament which was at that time very rare in Jutland.[21]

The third Borremose body was also female and aged twenty to thirty-five. She was well nourished and became known as the 'Borremose fat girl'. The body lay face down in the peat and was partly covered by a large woollen skirt. Her face and head were so badly beaten about that her features were unrecognizable. The initial conclusion was that she had been scalped and battered around the head. A second post-mortem conducted in 1977 revealed that there were no 'warding-off' injuries on her forearms (fractures acquired while shielding the head and face) so she had probably not been assaulted. Nor was there any trace of haemorrhage in the cranium, so scalping was unlikely. Haemorrhaging had probably occurred in the blood vessels of an eye and ear but had not permeated from the vessels into the tissues, indicating that the crushing of the head occurred after death.[22]

In 1950 peat-diggers unearthed Tollund Man – described as the face of prehistoric northwest Europe[23] – the most famous bog body in the world, admired for his extraordinary serenity and composure in death. With a height of 1.61 m (5 ft 3 in) he was short for a man of his time and, aged about forty to fifty, he is also relatively old for a bog victim. Beneath his sheepskin cap, his hair was closely cropped whilst the bristles of his stubbly chin can be clearly seen, so good is the state of preservation. Other than his hat he wore only a simple skin belt. The only other item found with him, still lying around his neck, was the long cord of plaited leather with which he had been hanged. From the lack of displacement of the neck vertebrae, it appears that he died through slow strangulation rather than by a neck-breaking drop or jerk. A grave had been cut into the peat and his body had been carefully laid on its left side with the legs bent. It seems that his lifeless body was brought to this spot and placed in its sleeping position with the eyelids and mouth closed:[24] after his death, somebody clearly cared for him. He was not the first body to be

28. Rear view of the
Elling Woman. The long
hair has been twisted
and plaited in an
elaborate fashion.

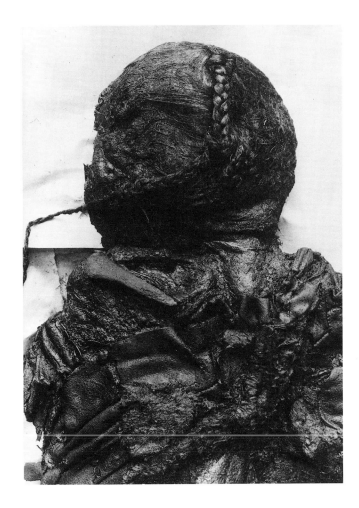

28. Rear view of the Elling Woman. The long hair has been twisted and plaited in an elaborate fashion.

found in the area. Only 70 m away had lain the body known as Elling Woman, who dates to the same period. She was about twenty-five to thirty years old and had also been hanged, in a slipknotted leather belt. Behind her very long and plaited hair the investigators could still see the V-shaped furrow this noose had made in the back of her neck.

The next Danish bog body to be found was Grauballe Man in 1952, a young man in his twenties with terrible teeth. Caries and peridontal disease must have given him toothache, while his wisdom teeth had never erupted and other teeth were missing. He also suffered from incipient arthritis in his spine. He too had been laid in a peat grave but his body was twisted and askew,

29. *Grauballe Man, one of the best preserved bog bodies from Denmark.*

unlike Tollund Man's. His throat had been slit from ear to ear – a 15-cm-long cut across his neck had severed his oesophagus. His left tibia was fractured and his skull was cracked from a hefty blow to his right temple. These injuries appear to have occurred around the time of death but we cannot say whether they were pre- or post-mortem. He was probably unconscious at the moment of his death. His intestine contents reveal that in the lead up to his murder, like the other Danish bog bodies, he had eaten a last meal of wheat and barley gruel which, as well as the usual weed seeds, also contained tiny splinters of bone. It seems he had eaten meat, perhaps pork, which may have been mixed into his soup. Like Tollund Man his food was contaminated with thousands of smut spores that had infected the barley. The bog moss in his intestine indicates that he, like Tollund Man, probably drank water from a bog pool, although it could have entered his stomach when he was put in the bog after his throat was cut.

In 1952 peat workers discovered the body of a girl at Windeby in Schleswig, in northern Germany. This find has been reinterpreted as a standard burial of the first century AD[25] but needs to be reconsidered. The body of this girl aged about fourteen lay in a dug grave, on her back with her legs slightly flexed, the left arm bent and the other by her side. She was naked except for a woollen band tied as a blindfold and a cowhide collar around her neck. Her hair was cut short and was shaved on the left side of her head. The body was weighted down with a large rock and covered with some birch branches and poles. There were no signs of violence, unlike Windeby Man, who was also placed in a grave and held down. There was a noose around his neck in the form of a hazel branch and he was pegged into the peat with many stakes.

The Windeby Girl is one of a surprising number of Iron Age teenagers who ended up buried in bogs. Another example is the Yde Girl from Holland who was aged about sixteen. She too had her hair partly shaved and was stabbed and strangled with a textile band which was wound three times around her neck, leaving a deep furrow in her skin. The Kayhausen Boy, from Germany, had his arms tied behind his back and was stabbed with a knife several times in the neck and left arm.

THE THREE BODIES FROM LINDOW MOSS

Of the three bodies from Lindow Moss in Cheshire, the most famous – Lindow II – was a man of twenty to twenty-five, clothed only in a fox fur armband, who had been hit twice on the head with a blunt instrument, at around the same time that he was garrotted and had his throat slit just above the ligature.[26] There was at first some disagreement among archaeologists about the latter two injuries. Steven Briggs claimed that the cutting of the throat was caused by later, more recent damage when the body was excavated, and that the garrotte was actually a necklace. The excavators pointed out that not only had this part of the neck been protected from any possible damage by the angle of Lindow II's head but also that the 'necklace' was tied in a slipknot thereby forming a noose.[27] The manner of this young man's death was dramatic and unpleasant. He was struck on the head so hard that bone chips from his skull lodged in his brain and the force of the blows shattered one of his molars. The types of insect found in the peat around the body indicate that the corpse was then immediately pushed down into a very wet area of bog. We do not know the sequence of actions but it has been suggested that the man was first subdued with the two blows and then garrotted and knifed in the throat. If the throat cutting was carried out while the garrotte was twisted tightly around his neck, then the insertion of the blade would have created a veritable fountain of blood spraying from the wound.

The fine condition of his fingernails indicates that he, like many of the Danish Iron Age bodies, was not involved in manual labour, at least in the months prior to his death.[28] So who was he? The radiocarbon dates indicate that he lived in the late first to second century AD, soon after the Roman armies occupied this part of Britain. Anne Ross and Don Robins have interpreted

30. The body of Lindow Man, with the front half cleared of peat. The body was cut through above the hips by peat-cutting machinery. Breakage of two of the vertebrae in the neck caused the head to rest at an abnormal angle to the shoulders.

31. *The sinew garrotte at the back of Lindow Man's neck. The garrotte may have been tightened as a tourniquet, but this was not the sole cause of his death: his throat had also been cut and he had sustained a massive injury on his skull.*

Lindow II's demise somewhat speculatively as the ritual death of an Irish Druidic prince.[29] They interpret the mistletoe pollen found in his stomach as evidence that he was a Druid (contrary to the more likely scenario that he was the 'sacrificee' rather than the sacrificer) and link the charred bread fragments also found to bread symbolism in eighteenth-century records of Scottish Beltain festivals (in which the person who receives the charred part of the bread is named as 'the devoted' and is subsequently referred to during the ceremony as 'dead'). Suffice to say, the circumstances of his death indicate a rite comparable to the mostly earlier Danish Iron Age bodies and best understood within a wider chronological and geographical context.[30]

The bog bodies in Lindow Moss came up in bits and pieces, having suffered serious damage from the peat-cutting machinery. Lindow II ('Lindow Man'), which was discovered in 1984, had been chopped through at the waist. His buttocks and left leg (originally recorded as Lindow IV) turned up on a nearby peat conveyer belt four years after the top of his body was discovered. The two other bodies provide something of the local context, in

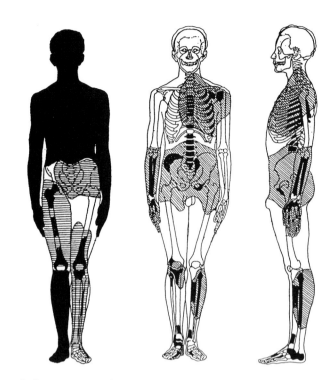

32. Probable positioning in body diagrams of the human remains from Lindow Moss.
Left diagram: *the parts of Lindow II (solid black) and Lindow IV (hatched).*
Centre and right diagrams: *the parts of Lindow III.*

spite of their incompleteness. Lindow I ('Lindow Woman'), was found about 250 m from Lindow II – although it is just part of a head (skull vault, hair, skin, left eyeball and part of the brain), it was identified by a forensic pathologist as possibly the remains of a thirty- to fifty-year-old woman. Lindow III, discovered in 1987, was badly damaged and consists of many pieces of an adult male. To complicate matters, the head that is Lindow I may actually be part of this male body even though it was found 55 m from the other fragments of the corpse. In spite of the damage to Lindow III's body, one of the preserved pieces was part of his hand – it had six fingers, the sixth being a small additional thumb (a condition known as pre-axial polydactyly). Both he and Lindow II suffered from the internal parasites that are so common in the intestines of bog bodies from the Iron Age. Both had infestations of whipworm whilst Lindow II was also home to mawworm. The relative frequency of whipworm in bog body intestines throughout northern Europe at that time indicates that this mildly uncomfortable condition probably affected most of the population.

The final surprise from Lindow emerged after the main body of results was published. A group of scientists hit upon the possibility that the skin of these people might have been decorated in ways not detectable by the naked eye. The bodies showed no evidence of tattooing but another question could be asked: had their skins been painted and how might one find out? Electron probe X-ray microanalyses indicated that the dermis layers of both bodies contained abnormally high concentrations of aluminium, silica and copper.[31] Although the copper can be explained away by natural processes, the other trace elements cannot. They point to the probability that these bodies were decorated with copper-based pigments often enough for the minerals to accumulate deep within the skin. The most likely interpretation is that these two men regularly and over many years coated their skin with a clay paste whose copper content would have given it a greenish blue colouring. According to Julius Caesar, writing of his visit to Britain in 55 BC, '[A]ll the Britons stain themselves with *vitrum*, which gives a blue colour and a wilder appearance in battle.' Pyatt and his co-workers argue that this unsettles previous assumptions that ancient Britons covered themselves in a dye made from the woad plant, though the discovery of carbonized woad from an Iron Age settlement at Dragonby in Lincolnshire[32] makes it likely that both sources of body paint were used.

A CONFESSION OF MURDER

The discovery of the head known as Lindow I set in train a remarkable sequence of events. It was recovered at the commercial peat company's depot in 1983 and, soon after, a local man confessed to the police that he had murdered his wife twenty years earlier and buried her body in the Lindow peat bog. Both he and the police were then surprised when archaeologists identified the head as part of an ancient bog body; the radiocarbon date confirmed the archaeologists' identification. The head was about 2000 years old, not that of a murder victim of the 1960s. Unfortunately, the facial tissue was destroyed before the head's archaeological significance was realized and only the skull vault remains. Some people, such as the forensic anatomist Robert Connolly, were still not convinced that Lindow I was ancient, having erroneously thought that the radiocarbon sample of bone was contaminated by peatwater of ancient date (like all such samples, it was in fact pretreated by the radiocarbon laboratory to remove potential contaminants). Connolly then proposed an odd scenario to explain Lindow II: perhaps he was the victim of a serial killer who had been stalking the moors nearly 2000 years ago.[33] One can understand what influenced Connolly's deductions – he was working within his own discipline's twentieth-century view of violent death, with its dreadful history of serial killers, moors and murderers. The archaeologist's perspective is very different, with the knowledge of a multitude of possibilities and motivations for prehistoric killings. To work towards the truth, archaeologists have to consider many different scenarios and reconstruct the physical and social context of these and other murders. Context is everything. And no, the murdered wife's remains were never found!

33. Human skull fragment from an Iron Age votive deposition site at Fiskerton, Lincolnshire. The skull shows a chop mark resulting from a blow with a heavy metal blade.

One clue about the deposition of the Lindow bodies into a bog comes from some human bones found in the mud around a wooden causeway that leads into the River Witham in Lincolnshire in England. The causeway at Fiskerton was built in the fifth to fourth centuries BC by people who dropped many swords, spears, tools, pots and ornaments into the mud on the river's edge.[34] Amongst these finds was part of the skull of an adult male who had been struck on the back of the head with a heavy metal blade, probably a sword. The Fiskerton site is a good example of the practice of votive deposition in which martial, industrial and domestic items were thrown into water without the intention of retrieval. This practice was very ancient, not just at Fiskerton but throughout northern Europe, beginning in the Neolithic and ending only in the fourteenth to fifteenth centuries AD.[35] The man from Fiskerton, known as 'Fissured Fred' by the excavators, may well have been another of these offerings. There is much speculation as to whom or what these offerings were made. There are suggestions that they may have been made to the moon, to the spirit of the river (similar to the medieval myth of the Lady of the Lake), and at times of calendrical significance or on the deaths of the owners of those items. We are not sure whether some or all of these factors were significant but votive deposition was a widely distributed practice forming an important backdrop against which northern European bog bodies should be investigated.

LAST MEALS OF THE IRON AGE BOG BODIES

Average heights of people in Early Neolithic Denmark were about 15 cm (6 in) shorter than they are today, equivalent to the average during the Industrial Revolution 150 years ago (1.65 m for men and 1.53–1.54 m for women). People's average height rose in the Late Neolithic and Bronze Age to reach a high point in the Iron Age of 1.72 m (about 5 ft 8in) for men and 1.62 m (about 5 ft 4 in) for women – this was not to be attained again or exceeded until the twentieth century.[36] It is most likely that these fluctuations in stature are related to changing diet and living conditions. After the Neolithic, people seem to have lived rather better on average in the Bronze and Iron Ages than their successors did under feudalism and early capitalism in the last millennium.

Iron Age communities sustained themselves on mutton, beef, pork, wheat, barley and a variety of pulse crops, on the evidence of food residues from settlements in Denmark and elsewhere in northern Europe. Our most graphic insights into Iron Age diet have come, however, from the preserved internal organs of twelve bog bodies whose intestines hold traces of what these unfortunate individuals ate before they died. The last meal of the man from Tollund in Denmark was reconstructed for television in 1954. A gruel of cereals and weed seeds, together with sand and bog moss, was fed to archaeologists Glyn Daniel and Mortimer Wheeler who pronounced it 'dreadful'. In Tollund Man's gut, scientists had identified the remains of oats and barley mixed with seeds from a variety of plants, some simply the common weeds found in cultivated ground (pale persicaria/ redshank, black bindweed, field pansy) as well as others known today to be edible (flax, fat hen, corn spurrey and gold of pleasure). Similar meals all containing large proportions of these weeds had been consumed by Grauballe Man, Borremose Man, and Huldremose Woman.

These seed mixes are similar to more recent 'famine foods' or to maslin (feeds of mixed grain) fed to domestic animals. Given what we know about Iron Age crop processing, the weeds would normally have been 'waste' and therefore picked out of a harvested cereal crop but during a time of crop failure or other food shortage, they may have been left in deliberately or even gathered separately. Food shortages certainly occurred: several bog bodies, including one from the Netherlands and another from northern Germany, exhibit the signs of arrested growth in the form of Harris lines within the bone. These are normally caused by malnutrition, illness or both. The case for these meals being famine foods is certainly plausible but we should also remember that a last meal might well have been deliberately designed to be unpleasant and different from the norm.

Curiously, the intestine contents from Iron Age and Roman period bodies outside Denmark show lower proportions of weeds. The Dröbnitz Girl from Poland had eaten wild vegetables, peas, and a gruel of wheat flour and meat. Dätgen Man from Germany had dined on millet and wheat with a handful of weeds and deer hairs. Lindow Man from Cheshire in England (one of three from that location and known as Lindow II) had eaten a mix of wheat, oats, barley and weeds which had been served up as unleavened bread, perhaps together with some meat, judging by the animal hairs and possible connective tissue also found in his intestines. The barley was not the best – it had been infected with a fungus known as barley smut (*Ustilago hordei*).

There were also four pollen grains of mistletoe in Lindow Man's stomach, which intrigued the excavators since classical sources mention mistletoe as a significant plant among the Gauls. Pliny the Elder, writing at about the same time that Lindow Man died, says that mistletoe was the most sacred plant of the Druids, imparting fertility and acting as an antidote against all poisons (*Naturalis Historia* XVI, 250–51). He describes how a white-garbed priest used a golden sickle to cut mistletoe from an oak during a ceremony in which the moon was hailed, before making a sacrifice; in Pliny's example the sacrificial victims were two white bulls. Of course, the mistletoe pollen could have been part of a herbal remedy – it cures cramp and soothes the nerves. From what we know of the violent manner of his death, Lindow II might well have needed something to sedate him before the sacrifice took place.

The samples from Grauballe Man's intestines (see above, p. 64) were unusual in that they contained many spore capsules of ergot (*Claviceps purpurea*), a fungus which attacks rye and other species, especially during damp summers. Ergot contains chemical compounds that we know today as constituents of the powerful hallucinogenic drug, LSD. Eating it causes disorders to the nervous system, induces hallucinations, and also restricts the blood supply to the body's extremities, leading to intense burning sensations and ultimately to withering of the limbs and incurable gangrene. When taken in large doses it leads to coma and even death. In medieval Europe the condition was known as St Anthony's Fire, on account of the burning sensations in the hands and feet, and there are records of whole villages suffering from this condition after eating diseased rye bread. Grauballe Man had swallowed enough ergot to send him into a coma, if not to kill him. So why had he been fed this extraordinary meal? It may have been a simple mistake, as a result of which his hallucinating fellow villagers happened to pick on him and kill him. Alternatively, he had already been singled out for killing and, unable to face his impending death with dignity, was fed this meal as a compassionate gesture so that he would not die in terror. The strength of the dose suggests that he was intended to be comatose rather than simply hallucinating. No other bog bodies have yielded up ergot in their intestines and so this intriguing case remains unique.

Despite the exceptional soft tissue preservation that can occur in bog bodies, there are few cases where the timing of death can be established with certainty. It has been suggested that the bog bodies found at Tollund and Grauballe were of individuals who died and were buried in the winter, as the food remains revealed no evidence of fresh vegetables or fruits. However, most cereals and seeds have good storage properties and can therefore be consumed at any time of the year. The stomach contents of Zweeloo Woman from the Netherlands included blackberry pips, as well as the usual gruel of various cereals. Blackberries are less likely to have been a stored food resource, indicating that the woman's death may have occurred in the late summer or early autumn.

Who Were the 'Bog People' and Why Were They Chosen?

The age profile of Iron Age bog bodies is strongly at odds with natural mortality patterns, since juveniles and young adults dominate the picture. There is no significant difference in the proportions of men and women but there is a surprising number of Iron Age and Roman period individuals with noticeable deformities. Lindow III was one in a thousand with his extra thumb. The Yde Girl had curvature of the spine (idiopathic scoliosis) which affected her posture and gait. Another Dutch bog body, the woman from Zweeloo, had short forearms and short lower legs (a condition known as dyschondrosteosis) that would have

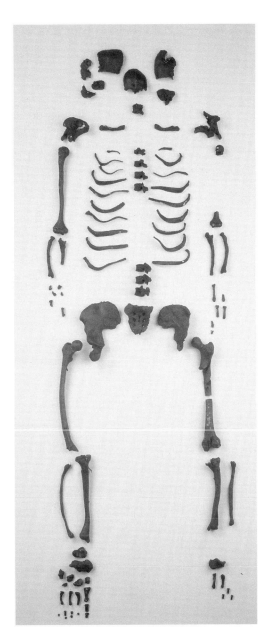

34. *The skeleton of Zweeloo Woman, found in 1951 and dating to the Roman period. The woman suffered from an unusual bone deformity called dyschondrosteosis, in which the bones of the lower parts of the arms and legs are abnormally short.*

given her an irregular gait. The man from Ashbroeken in Holland had not only suffered from malnutrition in his youth but had also broken his right arm. It had healed badly and would have stuck out at an unnatural angle. Finally, the Kayhausen Boy may have survived a festering infection in his right hip that would have prevented him from walking normally, though this diagnosis has been contested. John Magilton has proposed that these various deformities may have been significant, noting early medieval references to the unblemished physique of royal personages.[37] He suggests that people with minor physical defects may have been considered as 'touched by the gods' and therefore sacred, appropriate gifts to be returned to the supernatural world.

The incidence of deformities is surprisingly high, especially when considering the small number of bodies which have been adequately examined. However, the sample size is still inadequate to support any firm conclusions in this vein. Another aspect of these individuals' bodily presentation is that many show signs of having led a life free from manual labour, at least in the months leading up to their deaths. Hands are free of callouses or scars whilst fingernails are elegant and undamaged. The particular clarity of fingerprints may, however, be due to the preservative qualities of the peat (see Chapter 5, p. 162).[38] It is possible that a good number of these individuals were not ordinary farmers, a point borne out by the high-status artefacts associated with the women from Borremose and Huldremose. What can be said about the society in which they lived?

35. *Reconstruction of an Early Iron Age house at Moesgård, Denmark. These small houses provided living space for people and could accommodate a few animals over the winter. In the Late Iron Age more substantial longhouses were constructed with space for larger herds of animals.*

The Iron Age and Roman period in Jutland, the region from which so many bog bodies of this period are known, is very well researched. Many cemeteries, settlements and votive sites have been excavated and archaeologists have built up a surprisingly clear picture of social differences and changes.[39] During the first millennium BC and early first millennium AD, the period when most bog bodies were deposited, Iron Age societies in Jutland went through some dramatic changes.

The beginning of the bog body period, the Late Bronze Age, was a time of striking social differences, apparent in the diverse sizes of farmhouses and in the concentrations of wealth deposited as hoards of bronze artefacts. After about 600 BC, during the Early Iron Age, farmhouses were much smaller and most were approximately the same size though not all had stalling for keeping their animals over the winter. The extensive cemeteries indicate that most people (other than bog bodies) were treated at death in a consistent manner: after the body was cremated, the ashes were placed in a pot together with perhaps a single metal dress ornament, and then buried. As well as the bodies, other things were deposited in the bogs: simple or plain items (such as pottery, animal bones and foodstuffs), ploughs and bronze neckrings and even a remarkable boat complete with weapons. Early Iron Age dress fashions varied regionally with different ornament styles being worn in northern, central and southern Jutland.

After about 200–150 BC there were changes that ushered in an increasingly hierarchical and unequal society. On the margins

36. Iron Age votive deposits from Llyn Fawr, Wales. Many lakes, rivers and bogs throughout northern Europe were the sites of offerings of metalwork during the Iron Age.

of the three different regions of dress fashions, there sprung up communities who cremated their dead with the trappings of wealth – wagons, cauldrons, horses and cattle, weapons and imported metal drinking vessels – and lived in substantial long-houses with larger herds of animals than their neighbours. They were even using a black burnished pottery that differed from the brown pots normally used by everyone. By the first century AD this ruling class had established itself throughout the region; they now buried their dead without burning the body and no longer provided them with lots of grave goods, just occasional precious items of silver or gold. Relationships between men and women also changed, at least in how they were treated at death, with the sexes often buried in different areas or even in separate cemeteries. The goods now being put into bogs included fancy wagons, cauldrons and ornaments of precious metal.

Against this backdrop of social upheaval and change, bodies were put into bogs throughout. The relative imprecision of radio-carbon dating prevents us from linking individual deposits to particular historical moments so we cannot say whether periods of rapid and dramatic change were also times when more bodies ended up in bogs. The Borremose bodies date to around the transition from bronze to iron whereas Tollund and Elling date to the period around 200 BC. Grauballe may be a little later while the Huldremose Woman was buried in the first two centuries AD. It is not impossible that many bog bodies may have been victims caught up in revolutionary times. Some might have been mem-bers of a Bronze Age elite put to death in the new and more equal society that was ushered in with the adoption of iron. Others might have belonged to the incipient ruling classes that emerged after 200 BC and gradually consolidated their power in the follow-ing centuries. Until more precise methods of dating are available this theory of revolutionary execution cannot be dismissed.

Archaeological theories are often products of their time and cultural context. In 1937 Heinrich Himmler lectured his Waffen-SS officers on his own interpretation of the bog bodies and their moral message for Nazi Germany:

Homosexuals were drowned in swamps. The worthy professors who find these bodies in peat do not realise that in ninety out of a hundred cases they are looking at the remains of a homosexual who was drowned in a swamp along with his clothes and everything else. That was not a punishment but simply the termination of such an abnormal life.[40]

He was drawing on Tacitus' account of Germania, written in the first century AD, and referring to '*ignavos, imbelles et corpore infamis*' being pressed down under wicker hurdles and drowned in bogs and swamps. The link had first been made by another German, F. Arends, over a century before and was picked up by one of Himmler's SS officers, the archaeologist Herbert Jankuhn, who wrote a year later that male bog bodies had been cowards or had committed 'perverse sexual offences'. The passage is normally translated as 'cowards, shirkers and sodomites' but *corpore infamis* actually means 'disreputable in body' and could have a variety of meanings. Tacitus, of course, had his own axe to grind, using the Germanic peoples as exponents of a moral code with which to chastise what he saw as the degenerate lifestyles followed by his fellow citizens of Rome. He also tells us that traitors and deserters were hanged on trees and that adulterous wives were punished by their husbands, who cut off their hair, stripped them naked, turned them out of the house in the presence of kinsmen and flogged them all through the village.

The punishment theory is given further support from medieval documents. The *Lex Gundobada* of Burgundy rules that 'when a woman repudiates the man with whom she is united in wedlock, she shall die in a swamp', while the thirteenth-century *Edda* of the Norse peoples records the story of the lady Gudrun being accused by a servant of infidelity to her husband King Atle. The accusation leads to trial by cauldron in which the accused is found innocent and the accuser is found guilty and put to death in a swamp.[41] Finally, Elisabeth Munksgaard notes that, before the advent of the tumbril, Danish criminals were

The Wicker Image.

hauled to and from their place of execution on a cowhide so that they did not defile agricultural land by stepping on it.[42]

One of the problems with the punishment theory is that these historical sources, on which it relies heavily, relate to later periods. Even Tacitus' account, written at second or even third hand, is somewhat later than most of the dateable bodies. Tacitus has also been used to support another theory, that of human sacrifice. He refers three times to human sacrifice being practised by the Germanii, including the drowning of slaves in a secluded lake where the goddess Nerthus and her wagon reside. The sacrifice theory became popular in post-war Europe, fuelled by Glob's interest in rites of fertility and regeneration. Perhaps occasional human sacrifices were required in order to fulfil an obligation to repay the underworld for these gifts to the living, and the debt was repaid with that most precious votive offering that people could give – their own living kin. The case has been well made for bog bodies coinciding in the long term with depositions of food offerings in Danish bogs[43] and most archaeologists would support the sacrifice theory, bearing in mind that punishment executions might have sacred and sacrificial overtones. As Julius Caesar recorded of the Gauls, their gods preferred criminals but, if none were available, they sacrificed innocent people.

The crux of the sacrifice theory is that these watery places were also contexts for votive deposition of food, animals and artefacts sacrificed in order to make contact with the supernatural world. Thus it is through study of their contexts of deposition that we can begin to understand the motives behind the bog bodies' disposal. Recent excavations at the small raised bog of Almosen on Zeeland in Denmark have uncovered animal bones and pots dating from about 600 BC to the birth of Christ, together with bones from at least fifteen people, at least two of whom were children.[44] There are other sites like these in Denmark where the remains of adults and children have been recovered together with pots and food offerings. Yet, as Munksgaard points out, such places were probably open water in the Iron Age in contrast to

37. The Wicker Man, a giant wooden effigy that, according to Roman writers, was filled by the Celts with living sacrificial victims and burned.

the bogs where bog bodies have survived.[45] Those particular bogs which have yielded up bodies are generally not known for other votive finds such as pottery and animal bones. Borremose was used many centuries later for the deposition of the extra-ordinary Gundestrup cauldron made of silver but most Jutland bog bodies lie in isolation. Perhaps different places had different significance. Christian Fischer has suggested that many of the bog bodies are buried in areas where Iron Age people not only dug out peat for their fires but also searched for the phosphoric bog iron ore with which to make their tools and weapons.[46] Anyone who digs in peat soon discovers its preservative properties – did these Iron Age communities know that the bodies deposited in the bogs would not decay?

As the leading expert on bog bodies has written, 'different scholars hear different tales'.[47] It is too simplistic to characterize the diverse theories in purely political terms of 'punishment' = right wing, 'sacrifice' = ecological liberal, and 'revolutionary' = left wing, or in purely chronological terms of pre-war Nazism, post-war liberalism and 1960s/70s radicalism, though there may be certain links between interpretative and politico-moral stances. The evidence is often ambiguous, patchy and ill-recorded since many bodies were dug up long before archaeology became a scientific discipline. How we choose to interpret these bog bodies is not a matter of political leanings, though we should heed the moral abuses of the past. It is more a matter of detailed contextual study, through looking at landscape siting, artefactual associations in bogs and forensic details from individual cases.

CHAPTER 3

Mummified bodies

NATURAL AND ARTIFICIAL MUMMIES

*There are a set of men in Egypt who practise the art of
embalming, and make it their proper business.*
(Herodotus, *The Persian Wars*, Book 2)

Mummification is defined as the preservation of the soft
tissues of the body in a dehydrated state after death. Bodies can also be preserved by steeping in a preserving fluid (such
as the saturated environment of a bog) or by freezing, but in
these instances the tissues are not dehydrated and so bog
bodies and bodies preserved in ice are usually excluded from the
category of 'mummies' as conventionally defined. Typically,
mummified bodies are found in dry depositional contexts in which
the body is also protected from degradation by invertebrate and
vertebrate animals. As we have explained in the previous
chapters, mummification can arise from three distinct scenarios:[1]

1. *Spontaneous or natural mummification* can occur when a
   dead body is accidentally deposited in an environment in
   which the soft tissues become desiccated.

2. *Intentional natural mummification* results from the
   deliberate harnessing of the natural processes of
   desiccation by purposefully placing the body in a location
   that ensures mummification through natural means. In
   practice it is often difficult to distinguish spontaneous from
   intentional mummification, but contextual information
   may allow the investigator to postulate an intentional
   mortuary practice in some instances.

3. *Artificial mummification* refers to the situation in which direct human intervention is required in order to prevent the body from decaying. Usually some special treatment of the body, such as evisceration, the application of preserving or embalming materials or the enclosure of the body within a protecting structure, is regarded as evidence for artificial mummification.

We have also seen how the human body, while alive, protects itself against decay but following death is prone to immediate and irreversible decomposition. The principal agents of decay are micro-organisms that thrive under warm and moist conditions, and most of the world's population of today lives in temperate and moist tropical environments where decay proceeds rapidly. The earliest known mortuary rite, and still the most common today, is to bury the corpse in the ground, an action that in most environments results in skeletonization of the remains usually within a few years or decades. But throughout the world many communities have also buried their dead in more extreme environments such as waterlogged bogs, frozen Arctic soils and high-altitude sites, where bacterial growth is inhibited.

Hot and dry conditions also inhibit soft tissue decay as the tissues become dehydrated before decomposition occurs. The earliest evidence for deliberate mummification is, perhaps not surprisingly, found in regions of the world where the climate is hot and arid throughout the year; however, human ingenuity is such that in most of the climatic regions of the world some form of mummification can be and has been practised. Furthermore, the mummies that archaeologists study are those in which the soft tissues of the body have been preserved successfully for hundreds or thousands of years. There may have been many past cultures in which mummification was a temporary rite, with the corpse subsequently being treated in ways that render the period of mummification invisible to the future observer. In the nineteenth and twentieth centuries, anthropologists documented such practices in a number of societies.

38. *Grave houses of the Heiltsuk, near Bella Bella (Waglisla) in British Columbia. The style of the mortuary houses emulates the wooden-framed buildings introduced by European colonial settlers.*

The burial practices of the Heiltsuk (Bella Bella culture) of the northwest coast of British Columbia illustrate how mummification can result from a relatively minor level of intervention. Traditionally the Heiltsuk placed their dead in a flexed posture in a wooden coffin and the body then lay in state in the house of the deceased for a period of four days.[2] In some cases the corpse was then disembowelled and dried over a fire but often no special treatment of the cadaver was carried out. After the mourning period the coffin was transferred either to a special burial cave on a nearby island or to a cemetery, where it was placed in a purpose-built individual grave house (after European contact the preferred style of grave house became a miniature Victorian frame house, complete with windows and doors). The grave sites, and often the individual corpses, were tended and maintained for long periods of time, and preservation was assisted by the cold conditions that prevailed throughout most of the year.

At the same latitude as British Columbia, but 3000 km to the west, lie the Aleutian Islands which are strung in a great arc across the ocean between Alaska and Siberia. The maritime situation of the islands ensures that the climate is cold, damp and windy throughout the year. In spite of the harsh weather, the rich marine resources and extended habitable coastline allowed the native Aleuts to sustain a high population density. The Aleuts conducted burial rites that in some ways resembled those of the Heiltsuk: the Aleuts eviscerated their dead and then placed the body in a stream until it was clean. The body was then dried and the body cavities were stuffed with dried grass or *Sphagnum* moss, after which it was oiled and dressed in skin clothes. Finally the flexed body was wrapped in fine woven mats or skins of sea mammals and tied into a bundle. This was then placed carefully above floor level in a burial cave, together with supplies of clothing, weapons and tools.

In spite of the damp climate which would not seem to favour mummification the preservation of materials in the Aleutian mummy caves is surprisingly good. Wooden artefacts, nets, and baskets and mats woven from vegetable fibres have survived intact in the caves for hundreds of years. Organic materials are poorly preserved in outside archaeological sites on the same islands, and it appears that it is geothermal heat emanating from the rocks – because of the recent volcanic origins of the islands – that renders the air in the Aleutian caves unusually warm and dry, and thus conducive to the preservation of perishable materials.

The utilization of natural processes to assist mummification is also characteristic of the aboriginal funerary practices of the Torres Straits Islands, which lie in the seaway between New Guinea and the northern tip of Australia. A few days after death the dead body was taken a short distance out to sea in a canoe; here the outer layer of skin was stripped off, the internal organs were removed through an incision in the abdomen, and the brain was also removed through an incision at the base of the neck. After being returned to dry land, the corpse was tied to a wooden frame and allowed to dry in a vertical position in the hot air (Torres Straits mummies usually appear to have raised shoulders from the effects of the body sagging while being supported under the armpits). Puncture holes at the principal limb joints and in the hands and feet allowed the liquid products of putrefaction to drain out during the desiccation of the body. Special attention was given to the face: the lower jaw was tied to the skull with twine and after desiccation had been completed the appearance of the living face was restored by applying a surface coating of ochre with shells in place of the eyes. However, the mummification was not intended to preserve the corpse for more than a few years, and without further maintenance the mummy would eventually disintegrate.[3]

In South America we find the full spectrum of mummification processes, from spontaneous to artificial. Naturally desiccated cloth-wrapped bodies have been found in burial caves of the Guane Indians in the Mesa de los Santos region of northeastern Colombia,[4] and four naturally mummified bodies dating to between 4000 and 2000 BC were found in the Tres Ventanas cave in the Chilca valley of central Peru.[5] But the earliest mummies in South America are the elaborately prepared bodies of the Chinchorro people who occupied the Atacama desert region of southern Peru and northern Chile some 7000 years ago (see below).

Spontaneous mummification is less common in the more humid regions of the world, but as mentioned in Chapter 1, the Christian practice of placing the cadavers of important people in tombs inside ecclesiastical buildings often provided conditions conducive to preservation. In 1981 archaeologists excavating the site of a medieval abbey at St Bees in Cumbria discovered beneath the chancel aisle a tomb vault dating to about AD 1300.[6] The tomb contained the skeleton of a woman and a coffin enclosing the intact body of a man. The body had been covered in an embalming wax or resin, wrapped in a

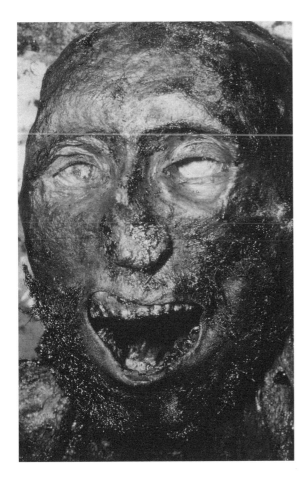

39. Preserved body of a man found at the Abbey of St Bees in Cumbria, England. The body had been embalmed, wrapped in a shroud and then placed inside both inner and outer coffins.

linen shroud, and then placed inside in a sheet of lead which served as an inner coffin. This had then been enclosed in an outer wooden coffin with the space between them packed with clay. Although the body had not been eviscerated, the embalming treatment, together with the enclosing layers of cloth, lead, clay and wood, had effectively prevented the body from decaying. When unwrapped the skin had a fresh pink appearance that rapidly discoloured on contact with the air. The eyes were well preserved (the colour of the irises could still be discerned) and the internal organs were all intact with the exception of the brain which had decomposed. One explanation for the unusual state of preservation is that this may be the body of a person who died abroad which was then embalmed so that it remained preserved during a long journey to the place of burial.

THE EARLIEST MUMMIES

The earliest mummies in the world are not, as one might have expected, from Egypt but from South America. At the river-mouths along the coast of the Atacama desert, in southern Peru and northern Chile, communities of fisher-gatherers known as the Chinchorro people were mummifying their dead as early as 5050 BC, some 2000 years before we find evidence of deliberate mummification in the land of the pharaohs. Perhaps what is strangest about these mummies is that the Chinchorro should have bothered to use artificial methods at all. The Atacama is one of the driest places in the world where rainfall is exceptionally rare. Natural mummification occurs here as a matter of course if a corpse is left buried in the sand. A second mystery surrounds how and why these people, living a simple and egalitarian lifestyle long before the advent of farming, developed such complex funerary practices.

40. Mummy of the Chinchorro culture, from the Atacama desert region of northern Chile, dating to c. 2500 BC.

Not only are the Chinchorro mummies the earliest in the world but they are also the most extreme in their modification. The process of breaking down the body and then remaking it is

unmatched anywhere else in the world. The earliest, after 7000 BC, were natural mummies wrapped in mats, furs or pelican skin. But by 5000 BC the Chinchorro people were using elaborate techniques to preserve the body. The head and limbs were cut off and flayed to keep the skin, scalp and face soft and supple for later. The brain was removed through the base of the skull and the internal organs were taken out of the chest cavity. Hands and feet were dried but the remainder of the skeleton was defleshed. Traces of burning on the bones suggest that hot coals or ashes were placed inside the torso cavity to cook or dry out the skeleton.[7] When the bones were clean and dry, the skeleton was rebuilt on a framework of body-length sticks reaching from the ankles to the neck, where the skull could be fixed on top. The sticks were then tied to the bones with mats and reeds. The skull and torso were stuffed with grass, ashes, soil, animal hair or a mixture of these, and then plastered and coated in a white paste made from ash. An artificial face – oval and flattened – with nose, eyes and open mouth, was built up on the skull from this white ash paste. This paste also formed the rebuilt limbs and torso with moulded breasts and genitalia as required. The previously removed facial skin was then stretched over the paste 'flesh' and the scalped hair was glued on top, together with pieces of sea lion skin. Finally the face and body were painted with a black manganese paste and the mummy was dressed with a loincloth or skirt and pieces of sea lion skin.

These black mummies were made over a period of 2000 years until around 2800 BC when they were replaced by several new forms. These included the red mummies, made by cutting off the head and cutting into the body to remove internal organs and major muscles before drying it out with hot coals. The corpse was not disarticulated, as in the earlier tradition, but was kept intact, stuffed with materials as before and strengthened by sticks pushed inside the legs, arms and torso, tied together at the neck where the head could be fixed back on. The skull was covered in a wig of human hair which was plastered in place with black paste, and the facial bones were covered in dark red ochre or black paste, sometimes after a primary coat of white paste. The whole of the body apart from the face was then painted with a red paste. This method continued to be used until around 2000 BC and was one of three forms of artificial mummification practised at that time. Some of the red mummies were wrapped, and are known as bandage mummies. There were also mud-coated mummies, fire-dried with coals and placed in their graves where they were coated in a thick cement of mud and sand with a protein binder which may have been blood.

The stylization of the faces indicates that the Chinchorro were not interested in reproducing the precise characteristics of the once-living individuals. The incorporation of sea lion skin might also have been significant in categorizing these dead as 'were-sea lions', that is, part human and part sea lion. Howard Reid has suggested that they may have mummified their dead to assuage the grief that is so strong in small communities and to remember and celebrate them as ancestors in physical form, manifesting the existence of a parallel world of the dead and providing a gateway to this other world.[8] Yet there are good reasons for thinking that these practices were not part of an ancestral cult. Many of the mummies are those of juveniles and children who, because of their young ages, would not have reproduced and could therefore not be considered as ancestors. It would be more appropriate to envisage them as a category of the dead – probably only a portion of the total population. Rarely does any society have just a single form of treating the dead and the low proportion of mature adults in Chinchorro cemeteries suggests that most of these may have been disposed of in ways which have left no trace.

Over the last few years archaeologists have had to revise some of their ideas about the technical feats achieved by pre-farming communities. The Chinchorro's extreme technique of mummification is one and the construction of earthen monuments in Louisiana around 6000 BC and the erection of giant pine-tree posts in Britain around 8000 BC are others. There are also two very elaborate burials with sophisticated grave goods from as early as 24,000 BC at Sunghir near Moscow. The problem arises

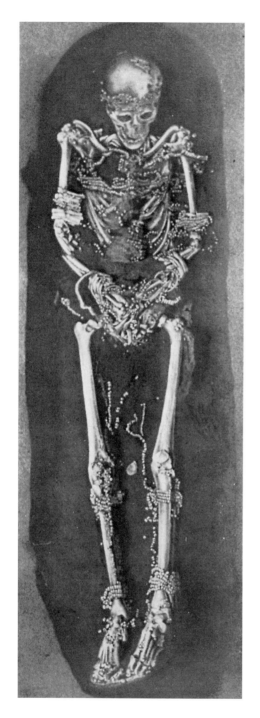 

41. *Elaborate Palaeolithic burials from Sunghir, near Moscow, from about 24,000 years ago. The beads ornamenting the adult male (left) can be seen wrapped around the skeleton. The double burial of two children (right) contains large ivory lances made from straightened mammoth tusks.*

from expectations that small-scale communities who subsist by hunting and gathering should be 'primitive' in their ideological and organizational capabilities. This notion developed in the nineteenth century out of European observations of other societies: they were seen as backward illustrations of stages along the road of progress from savagery to barbarism to civilization. This concept is untenable today but appears in archaeologists' perceptions of social evolution from 'simple' to 'complex' societies. Artificial mummification has been considered too complex for simple and apparently egalitarian pre-farming communities. And yet it clearly is not. Instead, archaeologists need to think of other forms of explanation which consider cases in context rather than as stages on an evolutionary scale.

Perversely it is through comparison with another place and another time that we can begin to understand the phenomenon of Chinchorro mummification. In the Near East around 7000 BC people were treating their dead in unusual ways. The most famous site where we have evidence of this is Jericho, the earliest walled town (as far as we know). From before 8000 BC people were removing the skulls from their dead, who were then buried beneath the houses and streets on which people walked. By 7000 BC many of these removed skulls were being modified in ways not wholly different to Chinchorro practices of 5000 BC and later. The facial bones were covered with a layer of plaster which was moulded and painted to exhibit facial features such as the nose, eyebrows and sometimes a faint trace of the mouth. The eyes might be formed by placing cowrie shells in the sockets. The skulls were then displayed by the living, as demonstrated by their discovery on house floors. It is unlikely that this was mummification in any sense because the bodies were buried and the skin and flesh presumably was left to rot before the skull was removed. The point of comparison between Jericho and the Chinchorro is not the making of a facial mask but the fact that the dead were physically present in both societies in a very direct sense. The Jericho skulls are, however, very different in age composition. All are from adults with the

42. *Decorated skull from Jericho, c. 7000 BC. The Jericho skulls were provided with faces modelled in plaster and eyes depicted using seashells as part of a cult of ancestor worship.*

exception of a single cache of infant skulls. Here it seems that adults of reproductive age were honoured in death whereas children and juveniles were not. They most probably represent a cult of the ancestors.

One of the interesting aspects of Jericho and other early farming communities in the Near East is the critical importance of the small oasis-like zones where crops could be grown in that arid landscape. People were demonstrating to themselves and to others inside and outside their community that their claims to these precious locales were very ancient indeed. They did so not only by displaying the skulls of their ancestors but also by constructing their mudbrick communities on top of their buried dead. These successive accumulations of occupation over centuries are known as 'tells'. Their heights were dramatic symbols of the long-term continuity and ancestral permanence of the residents.

In the case of the Chinchorro, their 'oases' were similarly precious and restricted. They needed the food resources of the sea and they also needed the freshwater of the rivers, along whose banks grew the vegetal resources which they gathered. Mummification was a means of bringing the dead physically into the community of the living in order to demonstrate timeless claims to place. The cemeteries are often found close to the houses where people lived and the evidence of repainting suggests that the mummies were displayed, possibly even standing up. Yet the approach of the Chinchorro varies in two interesting ways from that adopted in the Near East. Why treat all ages and not just adults? And why the extreme mummification when the desert could do the job for them? In the first place, ancestry itself may not have been important because the world of the living and the world of the dead were perhaps viewed as both separate and parallel domains. The dead were thus strangers, part sea lion and part human, who were separated emotionally and conceptually from the living. If the Eurasians of the Near East lived in a chain of being, counting back the ancestral generations and merging the gap between living and dead, then the South Americans were forgetting the dead as loved individuals and recasting them as a parallel community separated by an unbridgeable conceptual chasm. It seems that the extreme mummification, with its grisly defleshing, was necessary to dehumanize the dead in order to cross this chasm. Yet at the same time, by remaking them with attributes of the once-living and also giving them agency among the living it demonstrated a more encompassing attachment to place by both living and dead.

Finally, the remaking of these dead children into black and red mummies may have been an attempt to give them a second opportunity to remain in the world of the living. Although the mummies have been found buried in pits, they seem to have been kept for many years above ground, possibly on display, and several were restored by repainting and re-pasting their facial masks. Perhaps mummification was a way in which those whose lives were tragically shortened could enjoy their time on earth a little longer.

THE ORIGINS AND DEVELOPMENT OF MUMMIFICATION IN EGYPT

In one of the galleries in The British Museum one can visit a naturally mummified naked man known as 'Ginger'. He is from the Egyptian Predynastic period but there is a story, certainly untrue, that Ginger is not actually an ancient body at all but the corpse of a murder victim which was donated to The British Museum in the guise of an Egyptian mummy. No doubt the story originated because Ginger's condition of preservation is so good it seems impossible that he could be 6000 years old.

Ginger is typical of burials in the Predynastic period (*c.* 4500–3200 BC): he lies in a contracted foetal position with his hands by his face and was provided with a range of pots, tools and ornaments (the arrangement shown in fig. 43, however, is not original, since these goods did not come from his grave). He is so well preserved because the dry sand in his grave soaked up the decomposing bodily fluids until his corpse was entirely desiccated. People sometimes use the analogy of blotting paper and it is worth remembering that the first blotting paper in England was in fact sand.

As we have suggested earlier, people have always been aware of their environments and have developed detailed understandings of the properties of the soils that they work, dig, plant and harvest. Many Egyptologists have recognized that people in Predynastic Egypt would have been aware of the sand's capacity to create natural mummies.[9] They assume that ancient

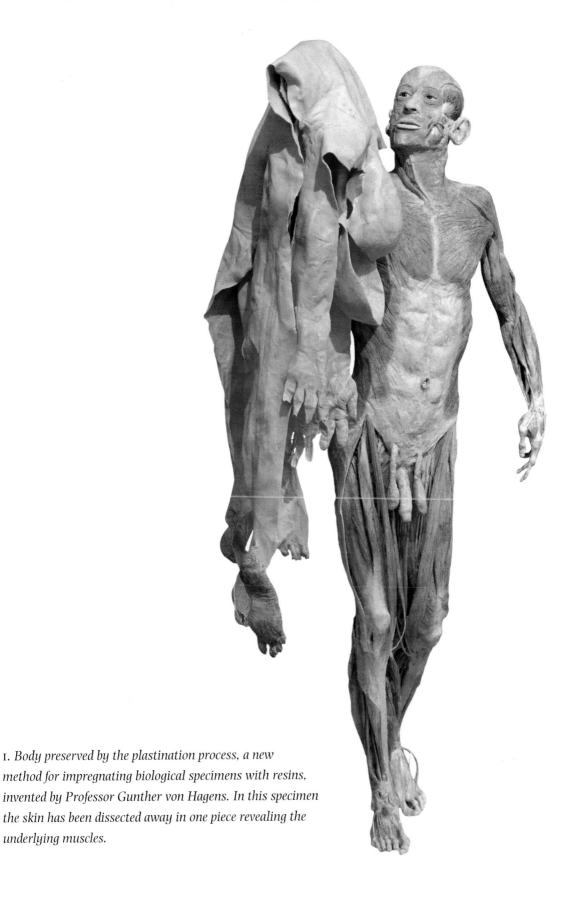

1. *Body preserved by the plastination process, a new method for impregnating biological specimens with resins, invented by Professor Gunther von Hagens. In this specimen the skin has been dissected away in one piece revealing the underlying muscles.*

II. *The body of Vladimir Ilich Lenin, displayed in his mausoleum in Moscow.*

III. *The exceptionally well-preserved face of Tollund Man appears peaceful, belying the violent manner of his death. Part of the plaited leather rope used to strangle him is visible around his neck.*

iv. *The body of a fourteen-year-old girl from Windeby, north Germany. The body lay in a grave in the peat with the limbs slightly flexed, and was naked except for a leather collar and a woollen band tied as a blindfold. The hair was shaved from the left side of her head at the time of death.*

v. *Two adult and two infant mummies from a Chinchorro burial site in northern Chile, c. 2500 BC. The defleshed bodies of the mummies were reconstructed using an ash paste and were then covered with skin and burnished with a black manganese paint.*

VI. *Inca mummies from Peru. The bodies were bound by cords in a flexed position and sitting upright, and would originally have been covered by layers of textiles.*

VII. *Graeco-Roman period mummy from Hawara, AD 100–120. By this time the Egyptian mummification process was no longer practised with great care. Externally the mummy has elaborate and attractive bindings and a painted face panel, but inside the soft tissues are poorly preserved.*

ABOVE VIII. *The body of Ramesses II in his simple wooden coffin. The mummy was first unwrapped in Cairo in 1886. Traces of the tobacco plant were detected in the mummy when it was scientifically examined in Paris in 1976.*

IX. *The face of Franklin expedition crew member John Hartnell, emerging from the ice that had formed soon after his burial from water that had filled his coffin.*

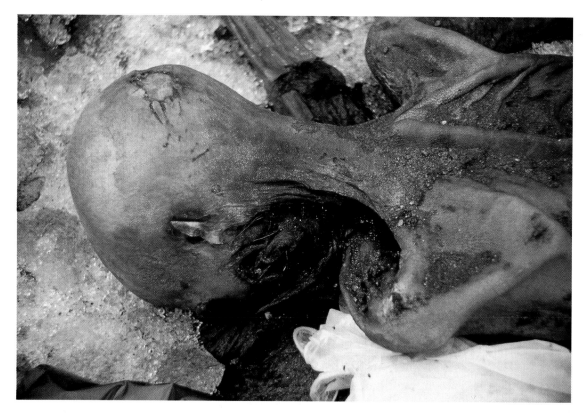

ABOVE X. *The Tyrolean Ice Man, photographed soon after he had been freed from the ice in which he had lain for 5000 years.*

XI. *The perfectly preserved natural mummy of an Inca girl, discovered in 1999 near the summit of the 6700-m-high (22,000-foot) Mount Llullaillaco on the border between Argentina and Chile.*

XII. *Cast of the surface of a human body preserved as a void within hardened volcanic ash at Pompeii. The skull bones are exposed at the top of the head.*

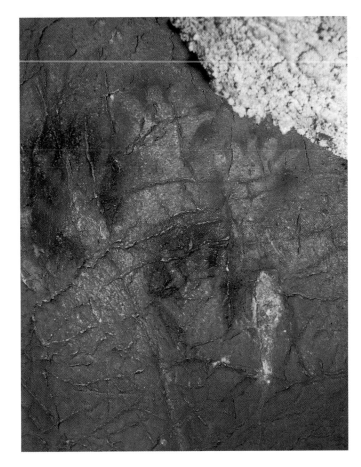

XIII. *25,000-year-old hand stencils, discovered on the walls of Cosquer Cave in France. The upper examples appear to have shortened fingers.*

XIV. *'The Atomic Apocalypse – Will Death Die?' A composition of over 100 papier mâché figures created by the Linares family in celebration of the Mexican Day of the Dead. The theme of the artwork is one of contemporary conflict and warfare.*

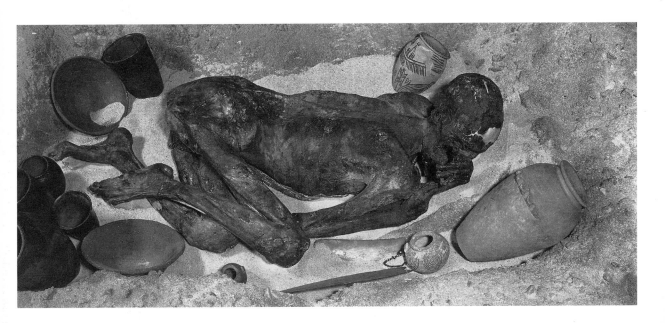

43. *Reconstructed burial of an unidentified adult man ('Ginger') of the late Predynastic period, from Gebelein, Egypt, c. 3400 BC. The bodies of Predynastic burials were naturally mummified by the dry burial environment.*

Egyptians would have known about this from occasional disturbances of earlier graves – by desert dogs or tomb-robbers – or perhaps from seeing forgotten graves exposed by the elements. From here, it is assumed, there was just a short and simple step to the development of artificial mummification of the kind that we know so well, first for the pharaohs and later for their courtiers, the common people and even animals.[10] Yet there are problems with this 'just so' story.

Surprisingly, the first efforts at mummification of the early pharaohs were not as successful as earlier natural mummification in the Predynastic period when bodies were left in the desert sand. Perhaps this was because preservation was required only for the duration of the royal funeral. Those performing the operation had either forgotten, or been entirely unaware of, or been unconcerned about, the preservative properties of sand. These early attempts are sometimes regarded as failures, put down to 'trial and error' or to poor judgement on the part of the early embalmers, a temporary glitch on the road of progress to perfecting the technique. But if we think about the length of time involved before the Egyptians finally got it right – decades and centuries – it seems more likely that the technique of

44. *Wrapped body of a woman of the mid-Predynastic period, excavated in a cemetery of Hierakonpolis.*

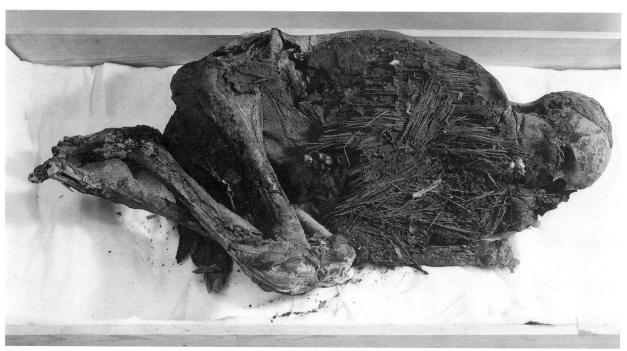

45. *A late Predynastic body from Gebelein, Egypt, originally placed in a wicker basket and covered in mammal skin.*

mummifying pharaohs was almost completely reinvented rather than copied directly from the natural world.

Exactly how artificial mummification started in Egypt is far from certain. The earliest evidence for wrapping mummies in linen and using resin and linen padding – at the head and hands only – comes from burials around 3500 BC in a Predynastic cemetery of Hierakonpolis.[11] These burials were not otherwise differentiated from those of others in this period so we cannot say whether they were of a special social status. So far only women have been found with wrapping and padding. Over the next 500 years, burials were sometimes wrapped in hides and matting and placed within small wooden chambers – removed from contact with the sand, the corpse and its wrappings decomposed. Paradoxically, another funerary rite at this time involved the deliberate dismembering of the corpse so that the flesh would rot and the bones could then be reassembled – often incorrectly – and wrapped as a skeleton bundle. Linen bandaging was used on corpses in the high-status graves of the First and Second Dynasties (*c.* 3150–2686 BC).

In 1899 Flinders Petrie began excavating a cemetery at Abydos, dating between 3150 and 2890 BC, the period known as the First Dynasty. He discovered a series of royal tombs, each surrounded by smaller graves containing the bodies of members of the court. Each of the royal graves had been robbed or ransacked and the bodies destroyed or stolen but, in the tomb of Djer, Petrie found a bandaged arm which had escaped the devastation. Only the radius and ulna survived but this arm was adorned with bracelets and amulets. Inside its bandage of fine linen, the flesh of the arm had not mummified because the body had not been in contact with sand. Unfortunately the bandaged arm does not survive today, having been thrown away by the then curator of the Cairo Museum who at least kept the jewellery. All we have of this evidence for First Dynasty attempts at mummification is a photograph and Petrie's notes. We cannot even be certain that the arm belonged to Djer. A French archaeologist, Emile Amélineau, who was responsible for some of the earlier excavation at Abydos, had found bodies wrapped in natron-impregnated linen (natron is a naturally occurring chemical, a hydrated carbonate of sodium). These could have been mummies of other First Dynasty pharaohs but they may have been much later in date.

The earliest evidence of near-successful attempts at mummification comes from the period of the Second Dynasty (*c.* 2890–2686 BC), from the great necropolis at Saqqara. These were not royal burials and the modest shaft tombs suggest that those buried here were not even members of the elite. Unlike later mummies, these bodies were laid out in a semi-flexed position. Whereas First Dynasty corpses had been wrapped in dry linen, these bodies had been plastered with linen soaked in a gummy resin which had moulded to the deceased's features, retaining the body's shape and appearance even after the flesh had decayed inside. The flesh had actually decomposed rapidly, creating a natural combustion that had charred the inner bandages of the mummies. The British archaeologist Walter Emery found the mummy of a woman whose breasts, nipples and genitals were moulded in perfect form. Each of her fingers was bandaged separately and carefully modelled.[12] Mere treatment of the body's surface was never going to preserve the body proper, and presumably the embalmers were aware of this. Instead, the practices of this period may have served merely to make the corpse outwardly presentable in the rites prior to entombment.

The pharaoh Djoser, early in the Third Dynasty (2686–2613 BC), is remembered as the builder of the first pyramid, the Step Pyramid at Saqqara. All that has survived in his tomb is a bandaged foot with its external features carefully modelled in the resin-soaked linen, lying amid the ransacked debris of his granite burial chamber. Radiocarbon dating indicates that these mummified remains are from centuries after Djoser's time and thus they belong to someone else.[13]

Although few mummies have survived from the Fourth Dynasty (2613–2494 BC) – the period of the great pyramids at

46. *Alabaster canopic chest allegedly containing the preserved internal organs of Hetepheres, a Fourth Dynasty queen from Giza, Egypt, c. 2575 BC. The viscera had been removed to inhibit decomposition of the body, an innovation first introduced at this time by the Egyptian embalmers.*

Giza – the provision in tombs of recesses and stone boxes for the viscera indicates that attempts were being made to inhibit decomposition by removing these soft internal organs. Tombs excavated at Meidum included these viscera recesses and that of Ranefer – a noble – still contained his linen-wrapped organs. His linen-covered mummy, its facial features enhanced by paint, was also recovered but disappeared in the London blitz. In the tomb of Queen Hetepheres, mother of Khufu – builder of the Great Pyramid – an alabaster box, divided into four compartments, contained natron solution and various organs wrapped in linen. Her body was never found and we cannot even be certain that these were her own body parts. During the Fourth Dynasty the mummies of the elite were laid out in the extended position and gradually other social classes followed suit. As was seen on the mummy of Ranefer, the eyes might be painted green, the mouth red and the hair black.

The Fourth Dynasty was thus a time of funerary innovations but there was still no means of actually preserving flesh beneath the resin-soaked bandages. This state of affairs continued for nearly 400 years, through the Fifth and Sixth Dynasties and the First Intermediate Period. It is not until the beginning of the Middle Kingdom (2134 BC) that there is evidence for the consistent use of natron as a means of desiccating the muscular tissue prior to the application of wrappings and resin. A group of Eleventh Dynasty princesses found at Deir el-Bahri had been dried in natron but their viscera were left in place – their unusually dilated anuses and vaginas suggest that a dissolving agent was injected through these orifices.

Why then did mummification become a suitable treatment for the pharaohs of ancient Egypt? Artificial preservation was certainly being attempted in the centuries before, although it was generally not successful. One suggestion is that the period of pre-burial treatment for these earliest pharaohs became ever longer.[14] As with Lenin's successors in the former Soviet Union, perhaps these new rulers felt that they had to legitimize their power and their succession by displaying the dead pharaoh's corpse for all to see and by building a splendid tomb whose lengthy construction would entail a delay before burial. Perhaps the embalming practices performed on those earliest pharaohs were never intended to preserve the body indefinitely but were merely a temporary measure to ensure the body's physical presence and survival during the lengthening funerary rites. We might well reflect on the ways in which embalming in the medieval and modern world has accompanied the cult of the leader employed by regimes bent on holding on to and extending autocratic and absolute power.

One of the difficulties in discussing Egyptian mummification is that it was a changing process, developed over a very long period of 3000 years. Its attendant myths probably also changed dramatically and in ways that were not always written down. A frequently debated question is whether the myths of pharaonic immortality grew out of the practice of mummification, or the practice out of the myths. Since the latter were only written down millennia later, we shall never know. Their most common version was eventually recorded by the Greek writer Plutarch around AD 100, in his book *Isis and Osiris*, as follows. The grand-children of the sun god Ra were the earth god Geb and the sky goddess Nut. Their children were Osiris (symbol of human resur-rection and ruler of the reborn dead), his wife Isis, his brother Seth and Seth's wife Nephthys. Seth killed Osiris and cut his body into small pieces, scattering them across the whole of Egypt. Osiris's soul departed to the land of the underworld, the Kingdom of the West below the setting sun. His sister-wife Isis then patiently gathered up every piece in order to restore him to life. But a fish, the Nile perch *Oxyrynchus*, had swallowed Osiris's penis and so he could never be made whole.

Plutarch's account tells us no more but other, Egyptian, texts continue the story: Isis was assisted by Anubis, sent down from the sky by Ra, who set about embalming Osiris's body by putting the pieces together and wrapping the reassembled body in its own skin. Isis and Nephthys managed to bring Osiris back to life briefly by fanning the body with their wings and Isis turned herself into a kite, hovering over him so that she might be impregnated with his sperm. Osiris returned to the underworld and his son Horus was able later to displace Seth as the ruler of Egypt and become the predecessor of the pharaohs.

These myths present us with the problem of anachronism. For example, Plutarch's account was a product of its time and place and cannot be taken as a transcription of a story that had remained unchanged for 3000 years. As contemporary studies of mythology demonstrate, myths change form and emphasis as the social and political circumstances of the myth-tellers and their audiences change. We have a similar problem with the methods used by the embalmers of ancient Egypt.

The most detailed account of the preparation of a dead body comes not from any Egyptian source but from the Greek historian Herodotus, that 'father of lies' or 'father of history' depending on your perspective. Curiously the many thousands of pictures that decorate the tombs and the texts of the *Book of the Dead* recovered from tombs and papyri are very coy about the process of embalming and mummification. It is the archaeological evidence on which we must rely, leavened by Herodotus' account written late in the history of mummification, in the fifth century BC.[15] Herodotus did actually visit Egypt and he was clearly fascinated by the practices, recording three methods of mummification:

1. The brain is removed through the nostrils using an iron hook and dissolving chemicals. The internal organs are removed through an incision in the abdomen cut with a 'stone of Ethiopia' (perhaps a flint or obsidian blade). The body is cleansed with palm wine, purified with incense, stuffed with

47. *Scene showing the ritual of the Opening of the Mouth performed on the mummy of Hunefer outside the tomb. From his* Book of the Dead *papyrus, c. 1310 BC.*

perfumes and sewn up. It is then 'soaked' in a bath of natron for seventy days after which it is washed and finally bandaged.

2. Cedar oil is injected through the anus and then the body is soaked in natron. The oil is removed from the body, carrying away the semi-dissolved internal organs. The body is cleansed and bandaged.

3. The body is washed and soaked in natron and then bandaged.

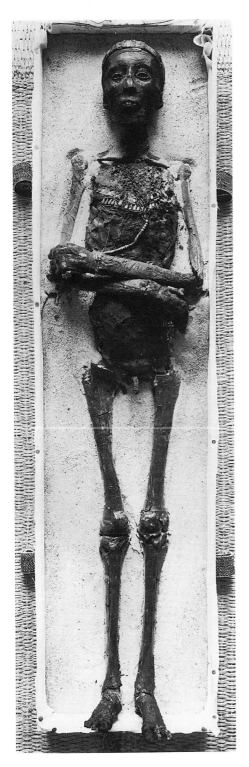

Between the Old Kingdom and later periods the appearance of the corpse was modified from a somewhat statue-like and idealized image of the deceased to the classic cocoon-like mummy-case within which were the wrapped body and limbs.[16]

From studying the bodies of the Middle Kingdom (2125–1650 BC) we can deduce that the process performed on them involved removal of the internal organs through an incision in the abdomen and removal of the brain through the nose, by rendering it into a mush with a probe. Natron was the principal substance involved in mummification. The packages of internal organs in Hetepheres' tomb were soaked in a 3% solution of natron but the preserving compound was most probably heaped around the corpse itself as dry salt. Tests on immersing animal corpses in a liquid 'bath' of natron solution proved that this method would have been ineffective.[17] A more recent experiment using dry natron on a modern human corpse donated to science has successfully replicated the degree of desiccation within seventy days.[18]

In amongst the bandages wrapped around the desiccated corpse were placed many small talismans and artefacts; in Tutankhamun's case the incision scar was covered with a gold plug. Unguents were poured onto the body during the bandaging. Fingers, toes, arms and legs were wrapped individually and then the whole body was bandaged. The unrobbed tomb of Tutankhamun

48. *The unwrapped mummy of the Eighteenth Dynasty pharaoh Tutankhamun. Many amulets and items of jewellery were enclosed between the successive layers of bandages wrapped around the mummy.*

and another from Tanis (of Psusennes I, *c.* 1039–991 BC) indicate that the pharaoh's face was enclosed within a golden mask. Tutankhamun's mummy was covered with his gold face mask and enclosed within three nested anthropoid coffins which lay within the stone sarcophagus.[19]

In the New Kingdom, between the Eighteenth and Twenty-first Dynasties (*c.* 1550–1069 BC) techniques and materials changed: a wider range of oils, resins and other materials were available; extraction of the brain became more common; and large quantities of resin were applied over the body.[20] Tuthmosis II (*c.* 1492–1479 BC) and later male pharaohs of the New Kingdom had their arms crossed. In the Twenty-first Dynasty, practices included the fitting of artificial eyeballs, and the subcutaneous packing of the body with mud and other materials. By the time that Herodotus and Plutarch were writing, in the fifth century BC and the first to second centuries AD, mummification was no longer practised with enormous care. The internal organs were left in place and the body's cavities and surfaces were filled and covered with resin. Externally the mummy looked pleasing with its attractive bindings and beautifully sculpted or painted face-mask but inside was an unpleasant mess.

These blackened mummies are probably the source of our use of the word 'mummy'. It derives from an Arabic word of Persian origin – *mumiya* – for bitumen, the black asphalt that covers our roads and for which Arab writers mistook the blackened resin coating of mummies.

COCAINE MUMMIES AND THE DIFFUSION DEBATE

For over a century researchers have drawn attention to some of the surprising similarities in techniques of mummification in different parts of the world, and some have argued that these similarities are evidence for the exchange of ideas between widely separated peoples through a process known as cultural diffusion. In the early part of the twentieth century

49. *Mummified adult male of the Guanche culture, from San Andrés, Santa Cruz de Tenerife. The Guanche mummification technique involved removal of the internal organs with the spaces being refilled with packing materials. The body was then washed, dried and wrapped with skins before being placed in a burial cave.*

the anatomist Grafton Elliot Smith and Warren R. Dawson (a former insurance salesman) studied the large collection of mummies recovered by the Archaeological Survey of Nubia. They drew parallels between the Egyptian mummification techniques and those used in the Canary Islands (Guanche culture) and Australasia. Elliot Smith explained the similarities by proposing that mummification was invented in Egypt and then spread to other parts of the world. Later research showed that both the techniques employed to prepare bodies and the rationale behind the practice of mummification in Australia and Melanesia were quite distinct from those in Egypt.[21] In spite of this knowledge, diffusionist theories are still popular and have been promoted enthusiastically by the Norwegian anthropologist Thor Heyerdahl, who has successfully navigated the Atlantic and Pacific oceans using boats constructed of papyrus reeds and balsa wood. Heyerdahl's research expeditions have shown that the papyrus boats that were used in the River Nile and the Mediterranean were capable of being sailed across the Atlantic Ocean to Central America.

Dramatic confirmation of ancient contacts between the Old World of Europe, Africa and Asia and the New World of the Americas appeared when Svetlana Balabanova and her colleagues at the University of Munich in Germany announced that they had detected traces of drugs in Egyptian mummies. By using forensic science methods the research group detected small amounts of drugs in samples of tissues taken from nine Egyptian mummies stored at the Munich museum.[22] Their results were unexpected and astonished the Egyptological world because traces of cocaine, cannabis and nicotine were present in all nine of the mummies. The problem that these results pose for archaeologists lies in the fact that cocaine and nicotine are active ingredients of the plants *Nicotiana tabacum* (tobacco) and *Erythroxylon coca* (coca) which are indigenous to South America and were unknown to the Western world until about 500 years ago. Nicotine is produced in small quantities by a few Old World plants but not in sufficient amounts to allow it to be extracted and ingested as a drug. The implication of Balabanova's work, if taken at face value, is that 2000 years before Christopher Columbus sailed to America, the ancient people of Egypt had access to plants or plant products growing only in the New World. Pollen and fragments of leaves from the tobacco plant were also detected in the abdominal cavity of the mummy of Ramesses II when it was scientifically examined in Paris in 1976, providing further evidence for contact between ancient Egypt and the New World.

There are several reasons to be cautious and sceptical about the 'cocaine mummy' evidence. Firstly, the chemical procedure that was used to detect cocaine and cannabis in the Egyptian mummies was radio immunoassay, a sensitive but not very specific technique that relies on the use of antibodies that recognize small parts of the molecule that is to be detected. Immunochemical methods can be problematic when applied to ancient materials, as cross-reaction can occur with a wide range of breakdown products from organic compounds that are not related to the target molecule, thereby generating false positive results.[23] Cross-reaction is particularly likely when polyclonal antibodies are used, as was the case in Balabanova's studies. Furthermore the Munich researchers did not perform any control tests, for example by analysing mummy wrappings or other non-human organic materials to see if these too might have tested positive for drugs.

A second major defect in the work is that Balabanova's group only tested for the active ingredients of the drug and they failed to report any tests for metabolites, the inactivated forms of the drug that are produced by metabolism within the body. For example, cocaine is broken down within the human body into benzoylecgonine, and detection of this metabolite indicates that the drug has been ingested and is not simply present by accident. Tests of the hair of Peruvian mummies that date to the twelfth and thirteenth centuries AD successfully demonstrated the presence of the metabolites of cocaine as well as the active drug itself, confirming that the ancient Peruvians had chewed coca leaves while they were alive.[24] By failing to test for metabolites in the Egyptian mummies, Balabanova's group cannot prove that the drugs were ingested by the individuals when they were alive. It is possible therefore that the mummies could have been contaminated by cocaine after they were excavated, either in Egypt or since their arrival in Germany.

The presence of nicotine in Egyptian mummies is even less likely to have resulted from ancient contact with the New World. The drug nicotine is known to act as a powerful insecticide, and this provides a straightforward explanation for the fragments of tobacco leaves and pollen which were detected in the mummy of Ramesses II. Museum curators formerly used powdered tobacco leaves and nicotine solutions to prevent moths and beetles from destroying organic specimens in their collections.[25] The Ramesses II mummy was partially unwrapped in Cairo in 1886, and it is likely that the tobacco fragments and nicotine found in this and other mummies were introduced by conservation attempts undertaken in Egypt in the late nineteenth century, soon after the original discovery of the mummies. Very small fragments of organic materials such as pollen grains can now be dated using the radiocarbon method, and it would therefore be easy to prove that the nicotine was introduced by the museum curators. Cocaine and cannabis are also known for their

insecticidal qualities and, before their criminalization in the twentieth century, were easily available for these uses.

CULINARY CLUES FOR KEEPING CORPSES

The symbolism of the cold meats was horribly obvious.
(Nigel Barley describing the meal eaten by the mourners after an English funeral)[26]

As we have already suggested, people's knowledge of how to mummify and preserve bodies may have derived from their observations of natural properties and processes. It is perhaps no surprise that artificial mummification has a long and ancient history within a region like Peru which is one of the driest parts of the world. That said, there are many desert regions where artificial mummification has never been practised. Equally, mummification has been practised in regions that are not especially dry. Preserving a body is more than a simple application of knowledge of natural processes. There has to be motive as well as opportunity.

50. *Gold mask and mummy bundle from a coastal Peruvian site of the Huari culture (Middle Horizon period, c. 800 AD). The body was seated upright in a tightly flexed posture on a coiled cotton mat before being wrapped in layers of textile.*

Ancient practices of body preservation are sometimes thought of in terms of primitive and misdirected proto-medical attempts to preserve a body in order to bring it back from the dead. Hollywood mummy movies trade on the fear that we feel in imagining that the ancient dead may actually come back to life. People tend to see mummification as not only a quasi-medical experiment on the road to technical mastery but also a complex once-only invention which those remarkable ancient Egyptians perfected and then spread to many different cultures all around the world. Such ideas are popular but they are also very wrong. Preservation of the body has come about in many different societies not as a single and original inspiration thought up out of the blue but as a practice grounded in other day-to-day activities.

From a purely scientific point of view, a dead body is just a

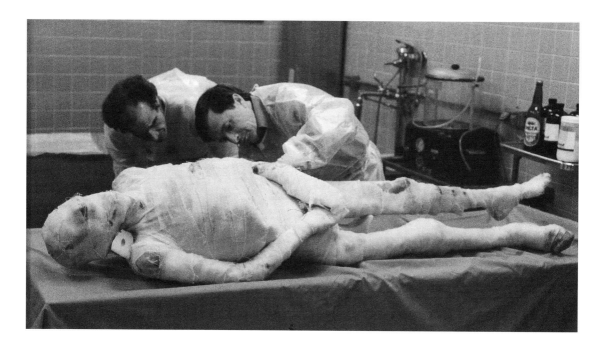

51. The American Egyptologist Robert Brier preparing a modern mummy from a donated cadaver, using the traditional methods of the Egyptian embalmers.

large piece of meat which soon begins to decay and gradually rots away. How to deal with this natural process has been a problem ever since humankind developed a consciousness that demands that we care in some way or other for the dead. From examining the graves of the very earliest humans, archaeologists reckon that this need may have come into being at any time between 4 million and 30,000 years ago, most probably at least 250,000 years ago.

When it comes to dead meat and skin, humans have become experts at preservation. We can smoke, sun-dry, oven-dry, pickle, salt, freeze, cool, tan, tin, bottle, cook, ferment and even control the rotting of the products so that their preservation is enhanced and decay is held at bay. The products can then be eaten, worn or used long after the animal in question has died. Some of these methods of slowing decomposition are very recent, the technical achievements of the last few hundred years. Others may be extremely old. Anthropologists working in different societies have discovered that people's local knowledge of such culinary methods may sometimes have a bearing on what processes they subject corpses to. For example, in parts of southeast Asia the decay of the corpse is considered to be linked metaphorically with a number of other transformations in everyday life in which decay is a necessary part of changing one thing into another: the manufacture of dye from rotting plant matter; the rotting down of hemp plants to make rope; and the fermenting of rice grains to make rice wine.[27]

Mummification in the Greek period in ancient Egypt was certainly conceived of as a metaphorical process – the Greek words for opening up and drying corpses during the mummification process are the same as those for gutting and curing fish.[28] There is even a pictorial representation of a corpse as a large fish in the tomb of Khabekhnet. The Nile perch – *Oxyrynchus* – plays a crucial part in the myths relating to the origins of death and fish were sacred to the goddess Isis who is also at the centre of these myths.

52. *Dried fish on sale in a market in North Africa. The gutting and drying of fish from the River Nile served as a metaphor for the transformation of the human body that took place during the mummification process.*

In medieval England and France embalming of noble or royal bodies was akin to pickling or cooking. Those who died abroad might be cut into pieces and boiled in wine or vinegar until the flesh separated from the bones. This method, known as *mos teutonicus*, was employed on the remains of Henry V whose cooked flesh, heaped with spices, was sealed with the bones in a lead case and shipped from Normandy to Westminster Abbey. The close similarity between preserving a dead body and preserving meat for consumption is strengthened by the term 'kitchens' for the side chapels of churches where this rite was carried out, and by the likelihood that butchers were involved in the process.[29]

In many societies, the prime symbols on which people draw to model and develop their own understandings of life and death are the basic processes of growing, living and dying that they see around them every day.[30] People throughout the island of Madagascar conceive of life as a process of hardening, with babies starting life as wet as water and the body gradually firming up into old age and death when it transforms into hard dry bones.[31] In parts of Papua New Guinea the living are often identified with a tree or a plant, as they grow, ripen, bear fruit and eventually turn into a dry and dead husk or shell. In parts of Peru and Bolivia, a region that was one of the great centres of mummification for millennia until the Spaniards arrived, the storage of foodstuffs has acted as a fertile seedbed of metaphors for processing the dead. Corpses are conceived of as being like potatoes: the desiccated husks are planted in November, the time of the Festival of the Dead, and bring forth new life in February, the time of carnival. The English expression 'gone to sleep' might be translated in another culture as 'turned into a potato'. Death is never simply death but is understood through a variety of symbols and metaphorical transformations.

BLOOD BROTHERS: THE EVIDENCE OF THE GENES

> PRINCE OF MOROCCO: *Mislike me not for my complexion, the shadowed livery of the burnished sun, to whom I am a neighbour and near bred. Bring me the fairest creature northward born, where Phoebus' fire scarce thaws the icicles, and let us make incision for your love, to prove whose blood is reddest, his or mine.*
>
> (William Shakespeare, *The Merchant of Venice*, Act II, Scene 2)

In mummified bodies the soft tissues may appear to be well preserved, but on a microscopic level the molecular constituents of the body are often altered during dehydration and by chemical processes which may continue long after death. Some molecules resist decay better than others, and amongst these are the blood group antigens which are the distinctive 'marker' molecules found mainly on the surfaces of blood cells but also in other parts of the body. Blood group antigens consist of polysaccharides, starch-like molecules which survive particularly well in dried tissues. Blood groups are inherited directly from the parents in a predictable fashion and they can therefore be used to test whether two people share a close genetic relationship. Harrison *et al.* demonstrated that the mummies of the Eighteenth Dynasty pharaohs Smenkhkare and Tutankhamun shared the same combination of blood groups, consistent with (though not proving) suggestions that the two men were brothers.[32]

Most blood groups occur quite frequently within populations, and therefore only limited conclusions can be drawn from their distribution amongst individuals. The best way to establish the genetic relationships between individuals and populations is to analyse the molecule deoxyribonucleic acid (DNA), the self-replicating genetic material contained within every cell of the human body. In 1984 the Swedish biologist Svante Pääbo reported the successful extraction and copying ('cloning') of strands of DNA from samples of skin tissues of two mummies from Egypt.[33] The DNA in the mummies had undergone some chemical change and many of these fragile molecules were broken into small fragments; nonetheless the results were still considered a scientific breakthrough. Prior to Pääbo's work many biochemists had believed that all the DNA in living organisms became degraded rapidly after death. Pääbo's discoveries heralded the dawn of a new discipline – biomolecular archaeology – which promises to revolutionize our understanding of the genetics of past populations.

Unfortunately there are still great difficulties involved in retrieving DNA from preserved bodies. DNA is only designed to function inside living cells, and after death the protective microenvironment of the cell is disrupted and the complex molecules of DNA are exposed to chemical decay. The principal factors that determine the rate at which DNA breaks down are the temperature of the burial environment and the presence of water. DNA is much more likely to survive after death if the tissues of the body become dehydrated or if the body is kept at a very low temperature. Pääbo's success with the Egyptian mummies was probably assisted by the rapid dehydration that occurred to the external surfaces of the bodies during the artificial mummification process, together with the dry environ- ment within the pharaonic tombs which prevented any subse- quent rehydration of the skin tissues. In contrast, the bog bodies of northern Europe contain no surviving DNA: the acid bog waters that are so effective in preserving the collagen fibres and hair of the bog bodies are in fact extremely damaging to DNA.[34]

DNA preservation is usually very good in tissues that have been preserved by freezing. However, the researchers who studied DNA in samples of tissue from the Tyrolean Ice Man dis- covered in 1991 (see Chapter 4, p. 129) were surprised to find that very little DNA had survived intact during the 5000 years that the Ice Man's body had lain in the Alpine glacier. They estimated that less than one millionth of the DNA that was originally pre- sent in the cells of the Ice Man's muscle, cartilage and bone had avoided degradation, and the small amount of DNA that was recoverable had been broken into short molecular fragments.[35] More recent biochemical studies of skin and muscle tissues from the Tyrolean Ice Man have shown that although the body had frozen soon after death it had later experienced a period of warming that lasted long enough for the body to thaw out and for the flesh to become colonized by bacteria.[36] It seems likely that the melting of the glacier that exposed the body in 1991 had occurred at least once before, and the damage to the Ice Man's DNA took place during one of these earlier melting episodes while the body was lying in a pool of water that had collected in the sheltered hollow where the body was eventually found.

Despite the promise of the early DNA results, surprisingly few successful studies of DNA in Egyptian mummies have since been reported. In part this may be due to the high average temperature in Egypt, as the likelihood of DNA survival decreases as the temperature at a burial site increases.[37] Lower average temperatures are found in the desert regions of northwest China, including the Taklimakan desert which has become famous for its remarkable natural mummies, often found complete with well-preserved clothing.[38] The exceptional aridity of the region together with the naturally high salinity of the soil ensures excellent preservation of buried organic materials. A cemetery at Qizilchoqa, near Hami (Qumul) in the Uyghur Autonomous Region, which was discovered by Chinese archaeologists in 1978, has been dated to 1200 BC. Over 100 mudbrick-lined graves have been excavated, and the bodies from this and other nearby cemeteries have distinctive facial features that are

53. The mummified remains of a woman from the Taklimakan desert, northwest China. The nostrils of the body have been stuffed with wool.

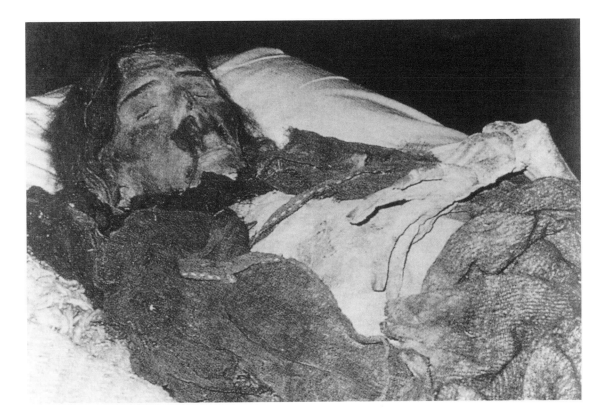

believed to indicate a European ('Caucasoid') rather than an East Asian ('Mongoloid') origin. Preliminary analysis of the DNA from one of the bodies by the Italian geneticist Paolo Francalacci has shown that it has a DNA type that is more common amongst present-day Europeans than in Asian populations.[39]

Thus it is only in exceptional circumstances that preserved bodies also contain well-preserved DNA. A variety of scenarios can lead to the intact survival of the soft tissues of a body, but in most cases the quality of preservation does not extend to the complex and delicate structure of the DNA molecule.

CHAPTER 4

Frozen bodies

ADVENTURE AND EXPLORATION: THE CHALLENGE OF THE UNKNOWN

To explore strange new worlds, to seek out new life and new civilizations, to boldly go where no one has gone before.
(Mission statement, Starship Enterprise)

Adventure and the exploration of unknown regions of the world are fundamental to the human spirit, and are celebrated in myths and heroic narratives in all cultures. No single period can lay claim to being the era of exploration. Humankind has been expanding the limits of the known world since hominids first migrated from Africa more than a million years ago. As early as 50,000 years ago groups using stone-age technology travelled across the open ocean to reach the world's largest island, Australia, and not long after the Arctic regions of northern Eurasia began to be occupied, at least on a seasonal basis, by bands of well-equipped hunter-gatherers.[1]

The desire to explore one's environment and to investigate what lies around the corner or over the next hill is not unique to our species – many animals also have an instinct to migrate to new territories. The grass in the next field is always greener, for horses as well as for people. But the human instinct for exploration is not motivated solely by economic necessities. George Leigh Mallory, who died in 1924 just below the summit of Mount Everest, famously stated his reason for climbing the world's highest mountain: 'Because it is there.'

Of course it is a truism that history is written by those who succeed and survive and the stories of many who perished during exploration inevitably may never be told. George Mallory's

54. *Members of the 1924*
expedition to climb
Mount Everest. Colonel
Norton, the expedition
leader (third from left)
stands between Irvine
(left) *and Mallory*
(right), *the two climbers*
who died near the
summit of the mountain.

fate remained an enigma until 2 May 1999, seventy-five years after his death, when an expedition on Everest led by Eric Simonson discovered the frozen body of the climber at an altitude of 8100 m. Finds of abandoned climbing equipment showed that Mallory had reached an altitude of at least 8480 m, and possibly the summit of Everest itself at 8830 m, but he had fallen and suffered fatal injuries while descending the mountain together with his climbing partner Andrew Irvine. Historians continue to argue whether Mallory and Irvine achieved the summit of Everest but their valour and determination to reach the top is beyond dispute.

Occasionally archaeology provides the key to unlocking the mystery concerning the fate of those who set out on a voyage never to return. The ill-fated expedition led by Sir John Franklin in 1845 was one of several unsuccessful attempts to find the northwest passage, an ice-free seaway through the Arctic Ocean that would offer a sailing route connecting the Atlantic to the Pacific Oceans. There was a strong economic motive for the search as the Panama Canal had not yet been built and the lengthy journey to the Pacific Ocean around the southern tip of South America was hazardous and time-consuming. Previous expeditions to explore the Arctic coast of North America had foundered through unfamiliarity with survival techniques, with crew members having inadequate clothing to resist the extreme cold and expedition diets lacking the essential vitamins for long-term survival without access to fresh food.

With the support of 128 officers and crew aboard two specially prepared ships, Franklin's expedition was equipped with the latest technology, including preserved foods sealed in tin-coated steel canisters. After over-wintering at Beechey Island in Canada's Province of Nunavut, Franklin and his crew successfully navigated most of the rest of the route only to make a fatally wrong choice of direction at King William Island. Had he turned his ships to the southeast at this stage Franklin may well have negotiated the narrow ice-free channel that skirts the mainland coast to reach the open waters of the Pacific Ocean, but he chose prematurely to turn west, entering a maze of pack ice from which the ships were never to escape. Franklin and twenty-three of his crew died during the next two years, and the remaining members of the expedition abandoned the ice-bound ships in April 1848, only to die of starvation and illness while attempting to escape overland through the Arctic wilderness.

In 1984 and 1986 the Canadian forensic anthropologist Owen Beattie excavated the graves of John Torrington, John Hartnell and William Braine, three of the crew members of the Franklin expedition who had died during the first winter of 1845/6 at Beechey Island.[2] Expecting the bodies to be exceptionally well preserved in their frozen graves, Beattie intended to use scientific techniques to try to establish the cause of death of these crew members who had died before the ships became beset in the ice. The post-mortem examinations were performed at the grave site by thawing the bodies using warm water; after completing the work and taking a few samples of tissue for laboratory analysis the bodies were returned to the ground and the cairns marking the graves were restored to their original appearance.

The autopsies of the bodies of Torrington, Hartnell and Braine showed that they had been suffering from tuberculosis and pneumonia at the time of their deaths but an important additional clue to their cause of death was found by analysing samples of their bones and hair. All three individuals had high levels of lead that had accumulated in their tissues, probably as

55. *The body of Franklin expedition member John Torrington, exhumed from his frozen grave in 1984. Scientific studies of the bodies of Torrington and other crew members showed that lead poisoning may have contributed to their deaths.*

a result of consuming preserved food from lead-soldered cans.[3] Studies of the bones of other members of the Franklin expedition, discovered on King William Island, confirmed that their lead concentrations were so high that they must have been showing signs of lead poisoning.[4] These results have been challenged on the grounds that lead was ubiquitous in the environment of nineteenth-century Britain, and that the sailors would have ingested lead from the food and drink consumed throughout their lives prior to joining Franklin's expedition.[5] Follow-up studies, however, showed that the lead concentrations were much higher in the vertebrae and the calcaneus (heel bone) than in the skull or the shafts of the long bones.[6] The vertebrae and the calcaneus are formed of spongy bone that takes up lead from the blood much more quickly than the dense bone of the skull and long bones, showing that the members of the Franklin expedition had suffered from acute ingestion of lead some time after the start of their journey. It is likely that the rapid accumulation of lead accelerated their deaths by causing neurological damage as well as anaemia and digestive problems. The added burden of their poisoned food would have had a severely detrimental effect on the ability of the expedition members to survive under the harsh Arctic conditions.

LIVING IN A COLD CLIMATE: PRESERVED BODIES OF THE ARCTIC

Early Western travellers to the Arctic repeatedly failed to learn some vital survival lessons from the aboriginal inhabitants of that hostile and unforgiving environment. The native Inuit were, by polite English standards of the time, savages who consumed fresh blood and raw meat, and whose table manners were unrefined – they used a knife close to their lips to cut portions of meat gripped between their teeth. Yet uncooked flesh provides the only natural source of vitamin C in the Arctic environment. The Western explorers, once they had exhausted their limited supplies of fresh fruit and vegetables, often succumbed to vitamin C deficiency and some died of scurvy, a debilitating disease that the Inuit successfully avoided. And in sub-zero temperatures how better to cut freeze-dried strips of meat than to first soften in the mouth the portion to be consumed, before cutting away the remainder?

A glimpse of how well equipped the native people were for survival in the Arctic environment is provided by the remarkable preservation of 500-year-old Inuit graves at Qilakitsoq on the west coast of Greenland.[7] The graves were discovered in 1972 by two brothers, Hans and Jokum Grønvold, who were on a hunting trip near a long-abandoned Inuit settlement site. The Inuit burials consisted of the fully clothed bodies of six women and two children. They had been placed under a natural rock overhang which protected the graves from direct sunlight and the effects of rainfall and snow. The bodies were dressed in warm clothing as if prepared for travel outdoors – consistent with the traditional Inuit belief that the journey to the afterlife was long and cold. Their layers of close-fitting skin and fur clothing were perfect for Arctic conditions, combining effective heat insulation and waterproofing with freedom of movement and moisture dissipation, qualities that are hard to achieve even with present-day survival equipment that uses the latest synthetic fabrics.

The Qilakitsoq bodies were not frozen in ice but had become desiccated by cold dry air that circulated through the stones that had been placed over the sheltered burial site. As is common in cases of natural mummification, the best preserved parts of the bodies were those that had dehydrated rapidly: the hands and the

56. *A naturally preserved body of a six-month-old child, from Qilakitsoq, Greenland, buried c. 1475. A combination of low average temperature and dry air at the rock shelter burial site led to exceptional preservation of the Qilakitsoq mummies.*

57. Detail of the forehead of a tattooed female mummy from Qilakitsoq, Greenland, buried c. 1475. The tattoos were created by using a needle to draw a soot-blackened thread of sinew through the skin. The designs signified a woman's family, tribal identity and marital status, while also enhancing the attractiveness of the face.

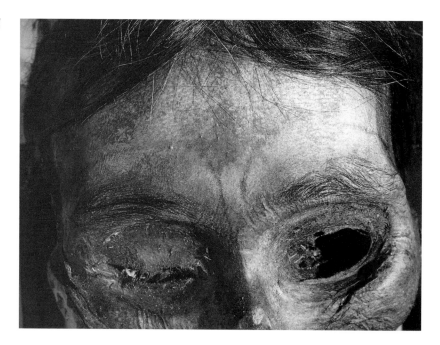

forearms. The most intact body was that of a six-month-old infant, again demonstrating that a baby's small body size is an aid to natural preservation. Tattoos were visible as dark lines on the foreheads of most of the women, indicating their status as married individuals, and radiography and autopsy of the bodies provided some additional information about their way of life. The adult women had extremely worn teeth, the result not just of chewing tough, partially cooked food but also because they used their teeth for preparing skins and other maintenance activities. Microscopic study of a sample of lung tissue from one body showed anthracosis or deposits of soot in the lungs, probably from inhaling smoke from burning seal fat which was used for cooking and lighting during the dark winter months. Some edible plant remains preserved in the intestines of one of the bodies indicated that death may have occurred during the short Arctic growing season at midsummer. There was, however, nothing to reveal the cause of death of these people; drowning, starvation and recent injury were excluded.

A number of other chance finds of frozen bodies have been made in the Arctic and these have usually provided invaluable information about past ways of life of the native inhabitants of the region. In 1982 Albert Dekin and Raymond Newell found the remains of a 500-year-old winter settlement of the Inupiat culture at Utqiagvik, near Point Barrow on the north coast of Alaska.[8] One house and its inhabitants had been catastrophically overwhelmed by surging pack ice, which had trapped and buried two Inupiat women and three children, together with many tools and household artefacts. The remains of the house were later flooded by summer meltwater, and when the water refroze it protected the house site from further erosion and damage. The frozen bodies of the two adult women were sufficiently intact to allow autopsies to be carried out which established that, as with the Qilakitsoq people, the Inupiat women had suffered from anthracosis from inhaling smoke inside their winter dwellings.[9]

An accident was also the probable cause of death of an elderly Inuit woman whose frozen body was found eroding from beach deposits at Kielegak Point on St Lawrence Island, which is located in the Bering Strait between Alaska and Russia.[10] The well-preserved body retained some hair and infrared photography revealed intricate tattoos on the arms and hands.[11] The body was unclothed, evidence that the woman had died inside a house rather than outside, and parts of her lungs were packed with fragments of moss which had caused some bleeding: she must have inhaled the moss as she was suffocated by the roofing material of a collapsing house. Radiocarbon dating established that this individual died around AD 400.

The publicity surrounding earlier finds of frozen bodies, especially after the sensational discovery of the Tyrolean Ice Man in 1991 (see below, p. 129), has ensured that newly revealed bodies are now investigated rapidly by anthropologists and pathologists who are experienced in the recovery and analysis of frozen remains. In August 1999 three Canadian schoolteachers on a hunting expedition in the Tatshenshini-Alsek Park in British Columbia found a body and artefacts in melting ice at the foot of a glacier.[12] The discoverers promptly informed the government

58. *Inuit winter clothing from Igloolik, Nunavut, Canada, 1986. The caribou skin clothing is perfectly designed for heat insulation, moisture dissipation and freedom of movement.*

heritage service and, with the cooperation of representatives of the indigenous nations, a team of specialists was recruited to record and retrieve the remains. The body was that of a native hunter who had fallen into a crevasse while crossing a glacier several centuries ago. His equipment included clothing of skins and woven fibres, and a collection of tools including walking staffs and a spear and atlatl (spear-thrower). A leather pouch containing filleted fish was also found alongside the body.

PRESERVED BODIES AT HIGH ALTITUDES

Outside the Arctic regions extremely cold conditions are found at high altitudes, and there is archaeological evidence that high mountain ranges were visited in prehistoric times. In September 1991 two German mountaineers were descending the Hauslabjoch glacier, in the Tyrolean Alps on the border of Austria and Italy, when they found a body emerging from a melting bank of snow and ice.[13] They reported the find to the authorities, who thought at first that the body was the victim of a twentieth-century climbing accident. Already that summer

59. *The Tyrolean Ice Man, discovered in melting glacier ice in September 1991. The extensive range of artefacts found alongside the Ice Man's body have provided a unique insight into prehistoric life.*

the melting glacier ice had released the bodies of five victims of modern Alpine mountaineering accidents. However, the artefacts found at the site quickly demonstrated that this body was of great antiquity and radiocarbon dating of samples of his skin and bone subsequently proved that the Ice Man had met his death between 5150 and 5350 years ago.

Anthropological studies showed that the Ice Man was in his thirties or forties and about 160 cm (5 ft 3 in) in height.[14] He was well equipped for travelling in the mountains, with effective wind- and weatherproof clothing, supplies of dried food and the means for lighting a fire. He also carried an extensive range of tools and weapons, together with materials for shaping stone and wood tools and for making running repairs to his equipment. It is clear that he possessed the means to be self-sufficient as a hunter, but some remains of cereals on his clothing suggest that he had stayed at a farming settlement not long before his death. Freshly shed pollen of the hop hornbeam tree (*Ostrya carpinifolia*) was recovered from his large intestine, indicating that he may have died in the early summer.

60. *Modern replica of the Tyrolean Ice Man's axe. The original axe was cast from almost pure copper, and the cutting edge was hardened by hammering.*

Analysis of the remains of bacteria in the body imply that the corpse had become frozen soon after death and this prevented the internal bacteria present at the time of death from causing the body to putrefy.[15] Much later, perhaps about 2000 years ago, an episode of climatic warming had caused the ice to melt and the body had then become desiccated on exposure to the mountain air. As is common with frozen bodies, the outer layer of skin (the epidermis), the hair and the fingernails had become detached from the body, but there were no signs of major putrefaction of the underlying muscles and connective tissues. The inner layer of skin (the dermis) was well preserved and showed the marks of tattoos along the man's spine and near his knee and ankle joints.

The cause of the Ice Man's death high in the mountains is not clear. He had several unhealed breaks in his ribs but these could have occurred long after death, either due to pressure on the body from the surrounding ice or during the extraction of the body from the ice. The researchers who studied the body have concluded that the Ice Man, perhaps weakened by injury and shortage of food, was overcome by adverse weather conditions while attempting to cross a high pass connecting two Alpine valleys. From the distribution of the artefacts around his final resting place it is suggested that some of the items were deliberately set aside before he rested in a sheltered gully on the side of the mountain.

A series of frozen bodies have been found in Inca burials near the summits of high mountains in the Andes in Peru, Chile and Argentina.[16] The Andes rise to over 6100 m (20,000 ft) and even today the high peaks present a formidable challenge to climbers equipped with modern technology. Yet 500 years ago the 6300-m (20,700-ft) summit of Nevada de Ampato was the site of an Inca sanctuary where children were sacrificed. In 1995 the archaeologist and mountaineer Johan Reinhard recovered the frozen bodies of three individuals from near the summit of this mountain: two were girls aged eight and fourteen, the other was a boy of about ten.[17] The children had been dressed in fine wool garments and the grave offerings included statuettes, pottery, wooden artefacts and corn. One of the children had been killed by a massive blow to the right temple, which fractured her skull and caused a fatal intracranial haemorrhage. Elsewhere in the Andes the frozen remains of at least another fifteen Inca sacrificial victims have been found, mostly at mountaintop sites above 5000 m.[18] The victims ranged in age from six years to young adults with a predominance of teenagers, and boys and girls are equally represented.

This form of sacrifice, known as *capac hucha*, required its child victims to be beautiful and unblemished, encapsulating the perfection, innocence and health of youth.[19] The ritual was considered to bond sacred space and ancestral time, and was performed at momentous transitions such as the death or accession of the Inca king or the beginning and end of the agricultural year. Children might be buried alive singly or in symbolically married pairs, and accompanied by votive figurines

61. *Mummified body of an Inca girl, discovered near the summit of Mount Llullaillaco on the border between Chile and Argentina.*

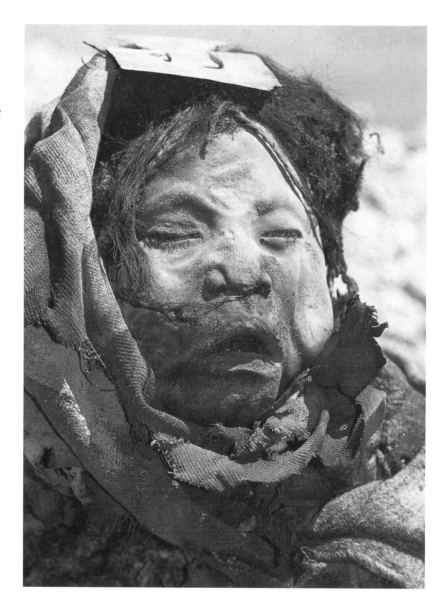

of gold and silver. The sacrifice was part of an elaborate formulation of the world and its workings in which the child victims were brought to the Inca capital and then taken to their place of sacrifice – a high mountain or an island or a volcano, for example – and offered to the sun god.

THE TATTOOED HORSERIDERS OF THE STEPPES

In 1924 – the year that Lenin's body was embalmed – another Russian, Sergei Rudenko, began some trial excavations of ancient burial mounds in the foothills of the Altai mountains on the border between the Soviet Union, China and Mongolia. In the valley of Pazyryk (which means 'burial mound' in the local language), an area inhabited by horseriding pastoralists, he came across a group of five large *kurgans*, round burial mounds. In 1929 Rudenko was able to excavate the first of these and discovered a wooden burial chamber from the Iron Age, built in about 400 BC and dug to a depth of 4 m. The most extraordinary circumstances had served to keep the chamber and its contents in deep freeze for nearly 2500 years. Every year in this region the frozen ground (known as permafrost) unfreezes during the short summer season, yet the stone and earth of the burial mound had created a microenvironment within the ground below, preventing it from ever defrosting. In addition, water or snowmelt had percolated into the tomb, to form a block of ice capped by permafrost. As well as non-perishable items, organic remains and any human and animal bodies would be perfectly preserved inside this unintentional freezer. To Rudenko's disappointment someone had been into the burial chamber before him and robbed it of the corpse and valuables. The bodies of ten horses, however, remained preserved and untouched.[20]

Rudenko realized that other mounds might hold similar if not more extraordinary finds but he was not able to come back for another eighteen years. The 1930s were difficult times for everyone in the Soviet Union, including academics, under the tyrannical and murderous grip of Stalin. War with Germany in the 1940s also prevented any archaeological research so it was not until 1947 that Rudenko returned to Pazyryk. During the next two years he excavated another four *kurgans* and this time he made some spectacular discoveries. Although all the tomb chambers had been entered and robbed centuries before, no doubt for their gold and silver, many wonderful items still

62. *Image of a horseman, depicted on a felt carpet from a burial mound at Pazyryk in the Altai mountains.*

remained. The tombs were full of leather, textile and wooden artefacts and each was provided with the complete bodies of between seven and fourteen horses with beautifully decorated bridles, harnesses, saddles and coats. In Barrow 2 at Pazyryk ice had prevented looters from robbing the entire chamber and here Rudenko found the preserved bodies of a man and a woman, although they had been pulled out of their coffin and now lay in pieces around it.

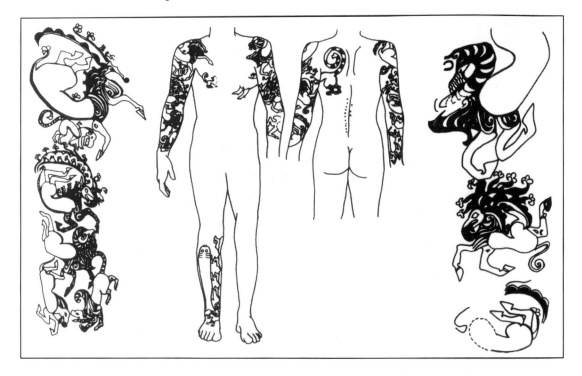

63. Distribution of tattoos on the body of the man in Barrow 2 at Pazyryk. Those on his right and left arms are shown in greater detail. There are no tattoos recorded on his left leg because the skin in this area was not preserved.

The man's body had been beautifully tattooed, with the forms of animals on his arms, legs and torso. The left hand and left leg had not been preserved but otherwise this was an exceptional find. The tattoos represented some very strange animals. Across the left side of his chest and curling around to his back was a winged lion, with a smaller lion on his right breast. Strange deer with eagles' beaks decorated his upper arms along with a monstrous clawed and fanged predator. On his lower arms predators (winged lion) and prey (rams and donkeys) appeared locked in struggle. There was also one of these strange predators and a line of rams, together with some sort of fish, on his right leg. The only non-pictorial tattooing was in the form of two lines of dots down either side of his spine.

The man had put on some weight around his midriff since he was tattooed, revealed by the way in which his body fat had developed between his tattooed skin and traces of the tattooing ink deeper in his flesh. He had died in late middle age, and was apparently scalped after being hit on the head with an edged weapon (possibly a battle-axe). What hair he did still have had been closely shaved but he had been provided with an artificial beard. A small leather bag at the head of the coffin contained fingernail clippings. Like all the other bodies found in these mounds, the tattooed man had been trepanned so that the brain could be taken out and the cranium filled with soil, pine needles and larch cones. His entrails had also been removed through an abdominal incision which was sewn up with sinew. The insides of his arms and legs had been slit along their lengths apparently for inserting some kind of fluid preservative into the muscles. Finally, his skin was pierced with small deep holes in his buttocks, legs and shoulders. Rudenko speculated that the preservative might have been salt but conditions within the tomb prevented any way of verifying this.

The woman in Barrow 2 was only partially preserved. She was over forty when she died and her hair had also been shaved – her plaits lay by her head. Given her placing in relation to the man, Rudenko considered that she was a concubine or junior wife buried with her master. She was evidently not tattooed and her body had been mummified in a different fashion to that used on the man. Her entrails had been removed through a slit afterwards sewn shut with horsehair, and the backs of her legs and buttocks had been cut open. The muscles here had been removed and replaced with a packing of sedge grass before the incisions were sewn up.

Bodies from Barrows 3 and 4 were preserved only as skeletons. There was a man and a woman in Barrow 5; he lay in the coffin on his back and she lay directly on top of him, also on her

64. *Coffin containing a preserved male body, from Barrow 5 at Pazyryk. The contents of burial mounds at Pazyryk were preserved by groundwater which had frozen after percolating into the burial chambers.*

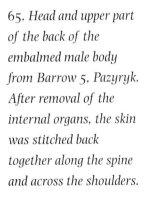

65. *Head and upper part of the back of the embalmed male body from Barrow 5, Pazyryk. After removal of the internal organs, the skin was stitched back together along the spine and across the shoulders.*

back. Both had been opened up more extensively than those in Barrow 2 so that their soft tissues might be removed. Their bodies were covered in long slits, down torso and limbs, front and back and the skin had then been stitched back together. The woman even had slits down the sides of her face. Horsehair padding had been stuffed into the woman's neck and breast area in order to preserve the body's shape. Curiously, the middle finger of the man's right hand was tied by a thread to the skin of his pubic area.

Aside from the tattoos, the most interesting aspect of these individuals' bodily treatment after death was the removal of soft tissue and subsequent repair of the invasive cuts into the body. Such observations would not have been possible if the corpses had rotted down to skeletons. These preserved bodies offer a fascinating, if at first sight inexplicable, insight into the death rites of these ancient horseriders of the high steppes. Why did they do it? The removal of internal organs is a practice that we recognize from ancient Egyptian and other mummifying societies – the bacteria in the gut and intestines cause this part of the body to putrefy fastest. Yet why remove the musculature, particularly by entry from the back?

The answer may come in part from observations by Herodotus, the same source as for some of our information about Egyptian mummification. Herodotus wrote about the burial practices of another group of steppe horseriders, the Scythians, who lived thousands of miles further west in the Black Sea region. He describes the funeral of a leader or king whose body had been carried about the kingdom on a wagon. The stomach cavity was slit open, cleaned out and packed with frankincense, cypress, and seeds of parsley and anise before being sewn up. The whole corpse was then covered with wax. Herodotus' account is, however, not without its problems: for example, can we really rely on evidence that was almost certainly not gathered at first hand?

The archaeology of the Scythians' distant neighbours at Pazyryk has revealed interesting corroborations that have led Rudenko and subsequent archaeologists to consider some of

Herodotus' information to be reasonably reliable. For example, Herodotus mentions the Scythians' use of cannabis smoke as a means of purification enjoyed at the close of a funeral, causing the mourners to become 'loudly exultant' as they inhaled the narcotic smoke. In each of the Pazyryk graves there were found the remains of small tent-like stands for burning and inhaling cannabis. In Barrow 2 there were two of these items of smoking paraphernalia: each had a small bronze vessel on which were piled seeds of *Cannabis sativa* which had been charred by hot stones placed on top of them. The resulting smoke would have been breathed in through the tent-like inhaler.

Between 1990 and 1995 another Russian archaeologist, Natalya Polosmak, continued excavations of more burial mounds and their frozen chambers.[21] In 1990 her first excavation

66. Preserved skull of a young woman who died 2500 years ago, from a burial mound at Ukok in the Altai mountains.

at Ukok, over 100 miles to the south of Rudenko's discoveries, revealed the skeletons of a man and a sixteen-year-old girl, both with weapons, but the burial had been looted at some time in the past. The next burial mound, dug in 1993, had never been touched. It seems robbers had found a later burial lying within its grave pit but had fortunately failed to notice that there was a primary grave below. Inside was the frozen body of a woman with tattoos, lying in a log coffin. She was accompanied by food offerings, leather and textile items, clothing, beads, gilded metal ornaments and a decorated hand-mirror. This 'frozen princess' had an unusually tall felt headdress – like a 'witch's hat' – causing speculation that she may have been a priestess though she was most probably just a member of the elite, distinguished from others by the presence of tattoos. Her mummified skin was only partially preserved, but tattoos adorned her shoulder, wrist and thumb. That on her wrist was of a stylized doe and on her shoulder was a deer with the beak of an eagle. The stomach contents of the six horses buried just outside her wooden chamber indicate that they died in spring, the season when the burial occurred. She was only about twenty-five when she died of unknown causes.

In 1995 Polosmak discovered another undisturbed burial, this time of a tattooed man – nicknamed the 'horseman' or 'warrior' – who seems to have died of wounds to the stomach. His hair,

67. Fragment of a tattoo on the arm of the frozen body of a woman found at Ukok in the Altai mountains.

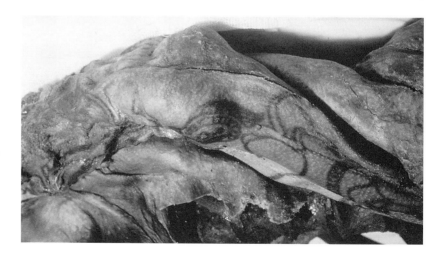

braided into two plaits, was still attached to his scalp and he wore a woollen cap, leather boots and a skin jacket of marmot and sheep. By his side lay his horse, together with his bow and arrows, axe and knife.

How can we explain these remarkable Pazyryk bodies? Rudenko reckoned that only the bodies of those of noble birth were embalmed. The removal of musculature was harder to explain and his suggestion that the flesh may have been eaten by the mourners is not wholly convincing. He also understood that another facet of life and death in the Altai region is the seasonality of frozen ground. Anyone dying in the winter months could not be buried until the summer sun had melted the icy ground sufficiently for a large grave to be dug. Preservation of a body over a winter would have been best effected by removing soft tissue to leave the skin stretched across the skeleton like a tent. While waiting for burial, such a prepared corpse would have had to accompany the living on their nomadic journeys.

There is perhaps a deeper level at which we need to understand the process of defleshing and evisceration. There are many ways to get rid of a corpse, and burial in the ground is just one of them. The number of trees used in the building of each tomb chamber at Pazyryk shows that there was sufficient timber available to have cremated all the bodies, regardless of the season. There are also many ways to display likenesses of the dead or the essence of a dead individual – the skeleton for example – without having to mummify the corpse. We must wonder whether the Pazyryk people were preparing the corpses in this way not only to survive the months before burial but to last indefinitely within the tomb. Like the Scythians, they lived in a world where, judging from their grave goods, felt and skins were enormously important. They surrounded themselves with decorated animal skins either ornamented with designs or cut into silhouette shapes – even the horses had leather face-masks. For a community on the move, their tents and material culture had to be easy to transport on horseback or wagon. They were expert at skinning and may even have employed metaphors of tents and frames in treating the dead.[22] The elaborate kitting out of each burial chamber indicates that they went into death well supplied with all the trappings necessary in the world to come.

Burials like the Pazyryk tombs and the Franklin expedition graves, or special finds like the Tyrolean Ice Man or the South American child sacrifices provide some of the most striking images in archaeology. At the same time, ancient deep-frozen corpses offer optimal circumstances for preservation and provide unparalleled opportunities for archaeological and forensic studies of clothing, hairstyle, tattoos, facial appearance, diseases, food remains and ancient DNA. Yet in may cases these finds, because of their extreme environmental settings, provide insights into highly unusual human circumstances rather than the conventional and everyday lives of past communities. The Pazyryk barrows are a rare example of the deep-frozen preservation of a society's normal burial practices.

CHAPTER 5

Impressions of the dead

THE SHADOWS OF DECAY: BODILY IMPRINTS OF 'SANDMEN'

> *There was no skeleton and no bone; the body had decayed to*
> *a hard dark brown crusty sand. With gentle trowel and brush,*
> *the line of the corpse could be seen or felt for – and so defined in*
> *three dimensions.*
> (Martin Carver describing the excavation of a 'sand-body' at
>   Sutton Hoo)[1]

Even when flesh and skin do not survive, the outlines of bodies can sometimes be recovered as stains in sandy soil. As the body rots in its grave, so sand invades the voids left by the dissolving tissue and becomes discoloured by the mineral products of bodily decay. In the right circumstances that discolouration of the sand where a body once lay may last for thousands of years. It takes a skilful eye to recognize these human body stains, which otherwise can be destroyed in an instant by careless excavation.

In August 1939, on the eve of the Second World War, a team of archaeologists excavated some of the most spectacular treasures of gold and silver ever found in Britain. They were found in the burial chamber of a ship buried under a mound at Sutton Hoo in Suffolk. Miraculously missed by ancient treasure hunters who had robbed other adjacent mounds but misjudged the centre of this one, the mound covered an extraordinary array of royal regalia, weapons, dress ornaments, bowls, plates and coins dating from about AD 620. Organic materials had largely perished except where their impressions had survived in corrosion products of iron and other metals. Some pieces of textiles and antler

68. Excavation in 1939 of the ship burial in Mound 1 at Sutton Hoo, Suffolk. *Organic materials at the site had decayed away, but the outlines of the ship's timbers were preserved as stains in the sandy soil.*

69. *Helmet from Mound 1 at Sutton Hoo. The helmet was found together with a sword, sceptre and metal clothing fittings but there was no evidence of a body which may have decayed leaving no trace.*

combs had been preserved fortuitously where they had been protected by being sandwiched between metal bowls. But otherwise the cloth, shoe leather, furs, bone and wooden objects, including the ship's timbers, had been destroyed by the acidic sandy soil.

A man's helmet, sword, jewelled epaulettes, purse, shoe buckles and sceptre lay within the former wooden chamber constructed amidships, but there was no sign of a body. Had it dissolved in the sand and its stained outline been rendered unrecognizable against the brown staining of the ship's timbers? Or was all that was left of the dead man the few crumbs of cremated bone that lay on a silver salver – whisked away by a gust of wind as the excavator carefully lifted the plate? At the time of the discovery the ship burial in Mound 1 was presumed to be a 'cenotaph', a symbolic interment of someone whose body had been lost or was otherwise unavailable for burial here. Many years later, the mound was excavated again and phosphate samples were taken from the floor of the boat to try, in vain, to discover if there was any trace of a body. Ever since, archaeologists have debated whether there ever was anyone placed within the Sutton Hoo ship.[2]

Nearly fifty years after the first excavation, archaeologists returned to the royal mound cemetery at Sutton Hoo under the direction of Martin Carver to investigate some of the other burial mounds and the spaces between them, and to develop new techniques of archaeological excavation. In one of the other mounds, Mound 2, they found another ship burial but this time the burial chamber was located beneath the remains of the boat, rather than within the hull of the boat as had been the case in Mound 1. It had also been robbed of any treasures and there was no trace of a body. Careful close-interval chemical sampling of the sand on the bottom of the chamber revealed at its west end unusually high concentrations of cations – aluminium, lanthanum, strontium and barium – the compounds indicative of a body. Their pattern suggested a human corpse was once placed here in a crouched position but this could not be demonstrated beyond doubt.

On the east side of the cemetery a series of flat graves was found in a group. The bodies that had been buried here survived only as outline stains in the sand. By using a handspray of 'vinamul', derived from polyethylene glycol (PEG), the excavators were able to trowel away the clean sand and consolidate the stained sand as they dug down into each grave. Gradually the corpse in each grave was revealed as a slightly amorphous body-shape of brown sand. The excavators christened them 'sandmen'. In some cases, teeth and bones survived. These bones lay inside the dark sand, indicating that this brown crust was the decayed remains of human flesh. A second group of sand-bodies

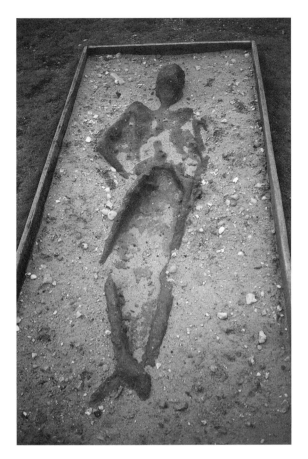

70. Body preserved as a stained outline in sand at Sutton Hoo. The stained sand, which results from the decay of the body's soft tissues, was consolidated with a chemical fixative during excavation.

was found in graves on the north side of the cemetery, encircling Mound 5. Some of these curious burials have been preserved on site as fibreglass castings made from silicon rubber moulds.

None of the sand-bodies had grave goods of metal or pottery and most were buried oddly. Some had been flung into their graves in twos or threes. Many were strangely contorted and twisted in different directions and one appeared to have had its neck broken. Another had been decapitated, the head placed on one shoulder. One body, nicknamed 'the ploughman', appeared to lie stretched in an uncomfortable hurdling posture astride what had once been a wooden object. Radiocarbon dates indicate that these burials are probably slightly later than the early seventh-century royal cemetery of mounds among which they were buried. They probably date to the eighth to eleventh centuries and Carver interprets them as the graves of people who had been publicly executed. But these were not ordinary criminals who had fallen victim to capital punishment. Carver views them as ritual killings of men and women who were at odds with the kingship, perhaps prisoners or just those who were somehow in opposition to the king. The wooden construction whose outlines were found in 'the ploughman's' grave he interprets as a dismantled gallows. There are even pairs of postholes in the eastern execution cemetery which may have supported the gallows' poles.

In the last season of excavation at Sutton Hoo, Carver's team discovered the grave of a man with his weapons, buried in a wooden chamber and accompanied by bowls and food offerings. Close to this grave was another, containing his horse and its bridle. After careful excavation, the outline of the man's body was easy to see and some of the bones survived. So what of the likelihood of a body in Mound 1? Had the pre-war archaeologists missed a body in the ship burial? Carver thinks so – he considers that the bottom of the ship and the burial chamber simply did not provide sufficient colour contrast for a body stain to be noticeable. We are not so sure. Items such as the shoebuckles and sword in the burial chamber of Mound 1 were not found laid out around the shape of a human body; they had not been arranged as one would expect had there actually been a corpse in the burial chamber, lying beside its sword. Unfortunately we will probably never know.

Elsewhere in the acidic sandy soils of eastern and southern England bodies are occasionally recognizable as dark stains.[3] This is also the case in parts of the Netherlands and Denmark where archaeologists have devised different ways of recording these sandy outlines. The bodies of people buried under Bronze Age barrows, or in flat cemeteries of Iron Age and early medieval date, can be recorded by photographing and drawing the outlines which are gradually revealed by removing thin strips or spits of sand across the grave. In this way a kind of tomographic slicing reveals the outline and disposition of the body. This method has the advantage of speed over the 'body sculpting' approach but provides less clear three-dimensional definition.

Risen from the Ashes: The Pompeii People

I like a girl with a proper mat, not depilated and shorn
Then you can snuggle in well from the cold, as an overcoat
she's worn.
(Roman graffiti from Pompeii)[4]

An unusual means of recovering absent bodies was presented to the excavators of Pompeii, the Roman town destroyed in a volcanic eruption from Mount Vesuvius on 24–25 August AD 79.

71. *Pompeii with Mount Vesuvius in the background. The cataclysmic eruption of Vesuvius in AD 79 killed many of the citizens of Pompeii, who died of asphyxiation and shock from exposure to the hot volcanic gases.*

The events which killed perhaps as many as 2000 people included five or six ash falls, of a type known by vulcanologists as *nuée ardente* ('burning cloud'), which buried the streets and houses, and the people who attempted to flee through the streets or stay in their houses. The corpses of those caught out in the open gradually decomposed within the fine and compacted ash so that the bones of the skeleton remained trapped in an empty space that had formerly been a body. Some of the inhabitants of Pompeii were thus preserved forever, but as voids, hollow impressions of their former selves. Of those who died inside the city's buildings, only the skeletons survive but their positioning and location still tell us much about these people's final predicament.[5]

We are lucky that Giuseppe Fiorelli, the director of excavations between 1860 and 1875, hit upon a way to restore these absent bodies to their physical forms. Each time that his workmen, shovelling out the ash layer, came across one of these voids they filled it with liquid plaster. Once the plaster solidified, the ash could be dug away from around it to reveal the contorted body of a victim of the volcano. We do not know just how many such voids were destroyed unknowingly before Fiorelli devised this simple technique. In the 1770s the impression of a female breast had been found and placed on display in a nearby museum but it was not until 1863 that Fiorelli had his ingenious idea. The method has also been used for recovering the shapes of garden tree roots (to determine what species were grown), wooden doors, furniture and fruit. More recently, researchers have used transparent resin instead of plaster in order to see the bones of the victims embedded within the casts. Fiorelli's lateral thinking has provided archaeologists with a unique view of Roman townspeople in the last desperate moments of their unlucky lives.

Many of the bodies display the so-called 'pugilistic attitude', in which the knees are slightly flexed, the forearms raised and the hands clenched into fists. This is seen today in bodies that have been recovered from burning buildings or vehicles, and it is caused by the coagulation of muscle proteins and consequent shortening of muscle fibres when the body is exposed to high temperatures. The white-hot volcanic ash from Mount Vesuvius burned the flexor muscles of the limbs and the exposed muscles on the bodies' backs, causing the arms to contract and the spine to extend as a result of the intense heat. The fineness of the ash is such that breathtaking details of facial features and clothing can be picked out on the plaster casts. With reference to the graffiti quoted above, the casts of several individuals are detailed enough to show that they had shaved their pubic hair into a semi-circle, a style also seen on several statues.

Perhaps one-fifth of the town's estimated population of 8000–12,000 died. Many perished in the narrow streets and we do not know how many more died in the fields and suburbs outside the town. A group who died at one of the town gates, the Porta Erculano, included a

72. *Cast of a head from Pompeii. The face was preserved as a negative impression in the hardened volcanic ash, allowing a replica to be made of plaster. The preserved skull of the individual is also visible.*

woman, her baby in her arms and two small girls clinging to her skirt. At the Porta di Nuceria (the Nucerian Gate) one man was found carrying a sack, and is sometimes presumed to have been a beggar. The impressions of sandals on his feet show that these were of good quality, leading archaeologists to interpret them as a gift from public charity. In the street nicknamed 'skeleton alley' three people were found who had died together: two girls and a large man, thought to be their servant or slave. The older girl lay on the ground with her face covered by her tunic.

Those who sheltered inside buildings suffered similarly violent deaths. One man and his dog had stayed put in the House of the Vestals; the dog survived long enough to start eating the man's body. One of the most famous casts is of another dog, its body in a contorted frenzy of agony, still chained up at the House of Vesonius Primus. It had climbed as high as its chain would allow through the falling ash before dying. More than sixty gladiators

73. *Cast of the body of a watchdog, House of Vesonius Primus, Pompeii.*

died in their barracks – together with a richly dressed woman whose presence there has caused much speculation. Two of the gladiators were locked in a cell and manacled at the wrists: like the dog, they had been left chained up to die in the turmoil. A valiant priest of the cult of Isis, running from the temple with the other priests, had become trapped in a house and, finding an axe, chopped his way through partition walls only to find his escape route blocked by volcanic ash. He collapsed with the axe still in his hand.

Recent vulcanological research has identified Vesuvius' likely sequence of eruption. It probably exploded during the early afternoon of 24 August, sending a giant cloud of ash high into the sky and turning day into night. After about half an hour pumice stones, formed from the froth of the magma, began to land in Pompeii and, after eighteen hours, had accumulated to a depth of almost 3 m across the town. During these first hours of the eruption some people, but not many, would have been killed by the falling pumice. We can only guess how many inhabitants had the good sense to take heed and abandon their houses for the safety of the countryside. Thousands certainly stayed put until the next morning when, one after the other, a sequence of six *nuées ardentes* formed pyroclastic surges and streams of hot ash which hurtled down the flank of the volcano and onto the town in a flowing and billowing torrent at up to 200 mph (300 kph).

As people were enveloped within the ash cloud they died from asphyxiation or suffocation, as a result of lack of air and the high toxic gas content, and from thermal shock as their bodies were roasted in temperatures of 100–400°c. The town was buried by these surges of ash and finally completely covered by the ash and pumice thrown up by a series of minor explosions which ended the eruption. The neighbouring town of Herculaneum was hit by a surge probably seven hours before the first reached Pompeii. It probably suffocated thousands of people and was closely followed by a flow of superheated rock and pumice. Unlike the situation at Pompeii, the people from Herculaneum are preserved only as skeletons and not as voids.

Characterization of the type of explosion that engulfed Pompeii is possible by examining the stratigraphy of the ash layers, the attitudes of the fallen, and an eyewitness account. The writer Pliny the Younger observed the events from a safe distance and wrote a letter recounting what he saw. He describes not only the sequence of the explosion but also the tragic fate of his uncle, Pliny the Elder, who was commander of the fleet at nearby Misenum and also author of that remarkable book on natural history, the *Naturalis Historia*. Pliny the Elder had seen the initial mushroom cloud from a distance and, inquisitive scholar that he was, got a boat ready to make a closer inspection. As he was leaving his house he received a note from a friend asking for rescue. He ordered the boat to head into the danger zone, from which everyone else was leaving, and landed at Stabiae, just south of Pompeii.

Here Pliny headed to the house of his friend Pomponianus, where he had a bath, dined and then slept, snoring loudly. Before going to bed he had pretended to his companions that the fiery lava flows were just bonfires or abandoned houses deliberately set alight! By morning the falling ash had reached such a height in the house's courtyard that the inhabitants were forced to decide whether to stay inside or to flee. They decided to run for it, amid the shaking buildings, and tied pillows on top of their heads to protect themselves from falling pumice and other objects. Pliny headed for the sea but there, engulfed by sulphurous fumes, he collapsed and died leaning on two slaves.

Pompeii is well known to archaeologists not just because it is such a complete survival of a Roman town but because the explosion's aftermath has provided the concept of the 'Pompeii premise'. When archaeologists dig up the houses and settlements where people once lived they are almost invariably looking at the residues of what was left after final abandonment. The remains that they find may have little to do with how that place was lived in during its heyday. One of the most important aspects of archaeology is, thus, the study of formation processes, or how remains have come to be as they are. Thanks to the suddenness of the eruption and the deep build-up of ash, Pompeii has, in contrast to most other sites, always been considered as a frozen moment – archaeologists can actually see just how life went on in this town and in its houses, without later disturbance. However, a second look at the events leading up to the final eruption – including the evacuation and the salvaging of precious objects – indicates that even Pompeii is not an example of the 'Pompeii premise'![6] The pyroclastic surges had not happened out of the blue or without warning: we know from Pliny's account that probably many thousands had already evacuated the town, taking many belongings with them.

INTIMATE MOMENTS: NATURALLY PRESERVED FOOTPRINTS AND FINGERPRINTS

> *It happened one day, about noon, going towards my boat, I was exceeding surprised with the print of a man's naked foot on the shore, which was very plain to be seen on the sand. I stood like one thunderstruck, or as if I had seen an apparition. I listened, I looked round me, but I could hear nothing, nor see anything; I went up to a rising ground to look farther; I went up the shore and down the shore, but it was all one; I could see no other impression but that one. I went to it again to see if there were any more, and to observe if it might not be my fancy; but there was no room for that, for there was exactly the print of a foot – toes, heel, and every part of a foot.*
> (Daniel Defoe, *Robinson Crusoe*, Chapter 11)

The fictional scene when Robinson Crusoe unexpectedly encounters a footprint on his deserted island is nonetheless a very real and familiar experience. The footprint in the sand is not merely an impression in an inert substrate – like a human shadow, it is a stark image that signals the imminent presence of another human being. Footprints are usually transient phenomena, easily eroded and obliterated by wind, water or the

74. *Footprints preserved in hardened volcanic ash at Laetoli, Tanzania. The footprints of three hominids are visible walking in a straight line away from the camera position: those of a juvenile on the left and of two adults on the right. The trail leading diagonally to the right of the hominid trails was made by a Hipparion (a fossil horse).*

movements of animals. In exceptional circumstances these erosional processes may not take place, and instead the footprints are preserved and hidden until they are uncovered by archaeological investigation.

In July 1978 Dr Paul Abell stumbled by chance on the oldest human footprint in the world, at Laetoli in northern Tanzania. The print that Abell discovered was just one of a series of footprints that were made 3.5 million years ago when a group of early human ancestors, members of the Pliocene hominid species *Australopithecus afarensis*, were walking across an open landscape that had been covered by a thick layer of ash from the eruption of a nearby volcano.[7] The hominid footprints were serendipitously preserved because the volcanic ash contained carbonatite, a mineral which hardens like cement when it becomes wet. After the hominids walked across the ash surface, a light shower of rain consolidated their footprints without washing away their details, and subsequent falls of volcanic ash then sealed the footprint layer. The whole sequence of sediments eventually became deeply buried and during the next 3.5 million years the layers of ash hardened into rock. The footprints were only exposed in modern times when natural erosion began to remove the softer overlying rock layers, revealing the slightly harder and more resistant surface on which the footprints are visible as foot-shaped indentations.

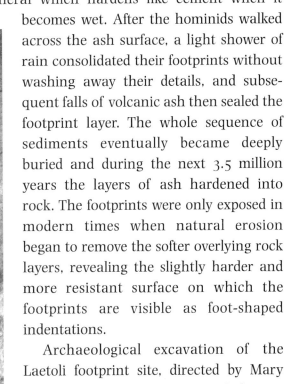

Archaeological excavation of the Laetoli footprint site, directed by Mary Leakey in 1978 and 1979, revealed sixty-nine footprints aligned in a group of trails running from south to north. The footprints had been made by three individuals who must have been travelling as a close-knit group. The right side trails were made by two individuals who had walked in line, with one leading and the other placing their feet in the leader's footprints. Just to the left of this first trail, a third individual left another trail of footprints. The separate footprint trails took a closely parallel course, with slight changes in direction mirrored in both the left- and the right-hand trails and a relatively constant distance was maintained between them. During the archaeological excavation the footprints were photographed and measured and the living statures of the individuals were calculated from the average length of the prints. In modern humans it has been established that foot length is about 15% of standing height, and it was argued that the feet were similarly proportional to stature in the hominids. The individual who made the left-hand trail was estimated to have been 115 cm tall (3 ft 9 in) and was therefore probably a juvenile. The two individuals on the right may have been an adult male and female, with statures of 176 cm (5 ft 9 in) and 141 cm (4 ft 8 in) respectively.[8]

The details preserved in the Laetoli footprints reveal the shape of the hominids' feet as well as their pattern of walking. The short distance between successive footsteps shows that the hominids were walking slowly, perhaps taking care as they negotiated the unfamiliar and possibly slippery terrain. Their feet made deep heel impressions and they gripped the soil between the ends of their toes and the balls of their feet, to gain extra purchase in the loose ash. The footprints show all the features of a modern human foot – the foot was strongly arched with the weight of the body transferring smoothly from the heel, through the lateral part of the foot, and then medially across the ball of the foot to the great toe. The slow pace of walking and the close grouping of the individual trails indicates that the hominids may even have been holding hands as they walked across the ash surface, perhaps in order to maintain their balance. This may be the earliest evidence that has yet been found for closely cooperative behaviour in a species that is ancestral to ourselves.

75. *Animal footprints at Laetoli, Tanzania. The large prints in the upper part of the picture were made by an elephant. The three footprints at bottom right may have been made by a hominid, but their outline is shorter and more splayed than the prints shown in fig. 74.*

By definition, footprints are made in soft unconsolidated sediments, and they are therefore only preserved in unusual circumstances. At the end of the last ice age, during the climatic transition from the cold Pleistocene to the temperate Holocene that took place 11,500 years ago, global temperatures increased rapidly and humans were quick to colonize the coastal regions of northern Europe. In the early Holocene the sea level was lower than in the present day because water was still frozen in the polar ice sheets, and many of the shoreline localities favoured by early Holocene hunter-gatherers were later submerged by the steadily rising sea. In a few locations around the coast of Britain, human and animal footprints are preserved in muds, silts and sands that were subsequently buried by fresh sediments as the sea level continued to rise.

At Uskmouth and Magor Pill, on the north coast of the Severn estuary in Wales, trails of footprints were made on silts that had been deposited near the shoreline by a high tide. Radiocarbon dating of the sediments showed that trails were made by Late Mesolithic people.[9] At Formby Point on the northwest coast of England footprints have been exposed along a 4-km stretch of the shore where coastal erosion is active at the present time. More than 100 separate trails of human footprints, together with numerous animal tracks, have been observed on shoreline deposits dating from the Neolithic to the Iron Age.[10] The trails are predominantly those of children and small adults (the latter probably women), and the slow pace of walking suggests that the individuals may have been collecting shellfish or other marine resources.

Some well-preserved examples of footprints of early humans have also been found deep in the interiors of caves. The presence of wet, fine-grained sediments in caves, coupled with the fact that caves are often secluded places which receive few visitors (either human or animal), ensures that once footprints have been created they have a better chance of preservation in caves than on open sites. Sometimes cave sediments become cemented by precipitated calcium carbonate, which serves to protect any footprints from erosion or damage by falling rocks. In several cases the footprints were made by children as, for example, the Upper Palaeolithic footprints preserved in caves at Aldène, Fontanet, Niaux and Réseau Clastres in France.[11]

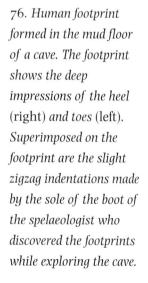

76. *Human footprint formed in the mud floor of a cave. The footprint shows the deep impressions of the heel* (right) *and toes* (left). *Superimposed on the footprint are the slight zigzag indentations made by the sole of the boot of the spelaeologist who discovered the footprints while exploring the cave.*

A LINGERING TOUCH:
TRACES OF FINGERPRINTS

Humans, monkeys and apes, and even rats and mice have ridged skin on the palms of the hands and the soles of the feet. These closely spaced sinuous ridges improve the grip when grasping objects (they are sometimes called 'friction skin') and also improve our sensation of texture. When we grasp a malleable plastic substance such as soft clay, the skin ridges form a negative impression in the flexible surface of the clay, resulting in a dermatoglyph – a fingerprint or palm-print.

In the nineteenth century Henry Faulds, an English doctor who was working at the time in a hospital in Tokyo, discovered dermatoglyphs on ancient Japanese pottery and he also suggested that ridge patterns might be observable on the preserved skin of mummies.[12] Faulds' main interest, however, was in the fact that in the fine detail of their ridge patterns each individual's fingerprints are unique; scientific fingerprint analysis subsequently developed as an important technique in the identification of criminals. The fingerprints that are recorded by modern law enforcement officers at scenes of crime are not in fact impressions – they consist of contact traces of sweat and sebaceous secretions that are transferred from the skin ridges to the surfaces of smooth objects. These organic fingerprints do not survive in the archaeological record, whereas physical impressions of hands and fingers on soft clays and on resins are much more durable and can survive for thousands of years.

The earliest evidence of a human fingerprint was detected on a small piece of resin found at the Middle Palaeolithic site of Königsaue in Sachsen-Anhalt, Germany.[13] Remarkably, the fragment of resin, measuring no more than 2.7 cm (just over 1 in) long, also preserved on one side the impression of a flaked stone tool and on the opposite side the wooden haft to which it had been attached, showing that the owner of the fingerprint had actually been using the resin to make a composite weapon or tool. Radiocarbon dating of the resin confirmed that it was between 45,000 and 52,000 years old, proving that it must have been used by a Neanderthal, long before the first modern *Homo sapiens* arrived in Europe.

Although natural resin preserves fingerprints well, it is an organic substance that does not often survive in archaeological contexts, so it is pottery and baked mud that provide the most frequent source of fingerprint records in antiquity. Fingerprints are best preserved on hand-made pottery, especially on areas such as inside the rim of a vessel, or where lugs or handles were attached to the sides of a pot by pressing and pinching the clay with the thumb or the index finger. Roughly formed domestic pottery is generally better for preserving fingerprints than is the glossy surface of fine ceramics: burnishing, glazing and the finishing of pottery by adding a slip (a coating of liquid clay) all tend to obscure any fingerprints that may originally have been present.

A large-scale study of finger- and palm-prints on Mediterranean pottery and clay tablets has been carried out by the Swedish archaeologist Paul Åström and his associates.[14] Åström's study of palmprints on Linear B clay tablets from Pylos in Greece has enabled the identification of the handiwork of ten individual craftspeople who were responsible for shaping the tablets before they were used for making written records. The clerks who inscribed the clay tablets scraped smooth the writing surface, but the unused sides and reverse faces of the tablets retained the handprints of their manufacturers.

77. *Fingerprints on modern hand-made pottery from Mexico. The surface bearing the fingerprints is on the inside of a globular vessel and was left unfinished.*

The identification of the handiwork of individual potters provides useful archaeological information that enables artefacts to be traced to a common source and proves their contemporaneity, at least within the working lifetime of an individual potter. In this way fingerprint identification can be used like refitting of flaked stone artefacts to establish stratigraphical relationships and to demonstrate common agency of production.

But fingerprint patterns can also be used to compare populations and to elucidate the population origins of groups of individuals. Studies of people living today have shown that there are significant geographical variations in the frequency of common ridge patterns such as the arch, loop and whorl patterns observable on the last joint of each finger. Fingerprints are genetically determined but as there are many genes involved, the pattern of inheritance is not a simple one.

A striking use for fingerprint studies is in the identification of the handiwork of children, who are otherwise often invisible in the archaeological record. Children's fingerprints are the same as those of adults, except in their size. In fact the detailed pattern of an individual's fingerprints is fully formed well before birth. The only changes that then occur between infancy and adulthood are a gradual enlargement of the print, with a widening of the spaces between adjacent ridges as the child grows in height and gains weight. The statistical correlation between the ridge spacing and the individual's age is remarkably high, and accurate measurements of the ridge spacing can therefore provide a fairly precise estimate of the age of the person who produced the fingerprint. Kathryn Kamp of Grinnell College in Iowa, USA, has studied fingerprints on clay animal figurines and domestic pottery from archaeological sites of the Sinagua, an ancient native people who occupied parts of northern Arizona. The figurines are small and often crudely made, and it is thought that they may have been created by children, while the pottery vessels from the same site were of high quality and were used for cooking and storage: by inference the pots would probably have been constructed by adults. The fingerprints on the figurines showed a significantly smaller average spacing between ridges and it was estimated that some figurines had been made by children as young as four, whereas the youngest makers of pots were at least ten years old and most of the pots had been made by individuals with adult-sized hands.[15]

Skin ridges have also been observed on the hands and feet of mummies and bog bodies. In bog bodies it is usual for the outer keratinized layer of skin – the epidermis – to slough off owing to prolonged immersion in the watery burial context, but the skin ridges are still visible, though with less relief, on the underlying surface of the dermis (you can sometimes observe the delicate dermal ridges on your own hand when you develop a friction blister, in which an accumulation of fluid separates the epidermis from the underlying dermis). The dermal prints on bog bodies, which are always delicate and unabraded in appearance, have sometimes misled researchers into treating this as part of the evidence that the bog people were of a high social status and took no part in manual work! The epidermis with complete skin ridges is more often preserved in mummies, and these have been photographed and replicated using dental impression material.[16]

Making an Impression: Handprints in Rock Art

Handprints are a common motif in rock art in several regions of the world. Handprints were produced either by dipping the hand in a coloured powder or liquid and then pressing it against a flat rock surface, or by placing the palm and spread fingers against the surface and then blowing or sputtering a pigment over and around the hand to produce an outline silhouette (a 'negative' imprint or stencil). Less frequently, pounding or engraving the rock surface created more permanent impressions of handprints and footprints. Many examples of stencilled handprints have been found in caves in Europe, Australasia and South America, and in some caves these handprints appear to show the mutilation or absence of one or more fingers. This has given rise to a variety of interpretations, from ritual body modification, to traumatic loss of the fingers through injury or frostbite, and even to suggestions that the cave painters suffered from the mutilating disease of leprosy. An alternative explanation for the missing fingers is that one or more fingers were flexed while the handprint was being created.

Fine examples of stencilled handprints have been discovered in Cosquer Cave near Marseilles in France. The cave is named after the diver Henri Cosquer who discovered the flooded entrance to the cave in 1985.[17] The cave entrance is now 37 m (120 ft) below sea level, but 20,000 years ago, during the last glacial period when the extension of the polar ice caps reduced sea levels far below those of the present day, the cave would have opened onto dry land and was accessible to the Palaeolithic hunters who occupied this region of Europe. Fortunately, the

78. Hand stencil in Cosquer Cave, France, dating to more than 20,000 years ago. The outline of the hand is cut through by a later incised engraving of a horse.

rising postglacial water level only flooded the lower half of the cave, and the walls and roof of the cave chambers above the water line are covered in paintings and engravings of animals and more than fifty stencilled handprints. The handprints are concentrated in the eastern part of the cave and are all of adult size and mainly of the left hand. They were created with red and black pigments and a number of unusual prints with an elongated forefinger may have been created by the same artist. Radiocarbon dates from the paintings and from pieces of charcoal preserved in the cave indicate two phases of activity, centred on 27,000 and 19,000 years ago, and direct dates from two of the hand stencils show that they belong to the earlier phase.

It is in Australia where ethnographic accounts of the production and meaning of rock art are most widely available. The insights obtained from the study of contemporary Aboriginal art can be used, with certain reservations, to help us interpret the prehistoric rock art of Europe and other regions of the world. According to the anthropologist Robert Layton there are two basic modes of representation in Australian rock art.[18] One tradition, which is expressed most fully in central Australia, depicts humans, animals and mythical beings according to the imprints that they leave in the sand, while the other tradition adopts a more figurative or representational style in which the external and sometimes the internal form of the creature is recognizable as a complete outline. Although stencilled handprints are partial rather than full outlines of the human body, they may have been individually recognizable amongst the members of the community and they are perhaps the equivalent of a modern calling card or even graffiti, a durable indication of the presence of a person at a special locality.

Interestingly, just like the Palaeolithic examples in Europe, some handprints in Aboriginal Australian and Tasmanian art display missing segments of fingers or even the absence of whole digits.[19] Although some observers have interpreted the stencils with missing fingers as evidence of deliberate mutilation or, less plausibly, of disfiguring diseases,[20] others have suggested that the

stencils may depict hand signals.[21] In many cultures the hands
are positioned in particular ways to signify identity, or to communicate using gestural symbols (familiar contemporary examples in Western cultures are the stylized hand gestures of urban
street gangs and the hand signals used for silent communication
by members of SWAT teams). By flexing one or more fingers the
stencilled handprints of rock art may communicate a message
that goes beyond simply indicating the individual's presence,
although if this does represent a gestural language the secret
messages of the Palaeolithic artists are still unknown. Handprints with missing fingers at the cave of Maltravieso in Spain
have been studied using ultraviolet light, which has revealed
them to be complete hand stencils in which one finger was subsequently painted out.[22] This reinforces the view that some of the
'mutilated' hand stencils in cave art may have been symbolic in
purpose, rather than being faithful representations of the
anatomy of the artist.

FACING THE PAST: PREHISTORIC BODILY REPRESENTATION

We will conclude our account of impressions of the dead by considering, albeit briefly, the origins of deliberate representation of
the human form through sculpture and painting. We have seen
how artificial mummification may have developed out of an aware-
ness of the naturally preserved dead, and how in those societies
that practised it mummification may have served to reinforce the
presence of the dead within or alongside the living. The earliest
mummies created by the Chinchorro were given artificial faces
that were sculpted from an ash paste, but the stylization of the
mummies' faces shows that the intention was not to reproduce a
physical likeness of the once-living individual. The mummies also
incorporated elements from the natural world, including parts of
other animals. These were the created dead, rather than the *re-
created* representatives of their living community.

One of the enigmas concerning Palaeolithic art in Europe is that (with the clear exception of handprints) depictions of the human form are uncommon and are usually less detailed and more abstract than the realistic and accurately portrayed images of ice age animals. Painted images of people are very rare and most of the examples of human figures that have been discovered are either outline drawings engraved on the smooth surfaces of bones or stones, or three-dimensional figures carved from ivory. Depictions of the human form are often stylized or schematic: the limbs are of disproportionate size and details such as the face, hair, hands and genitalia are usually absent. In contrast, paintings of animals are frequently life-like and well observed, and include accurate details of the eyes, the patterns of body markings, the animal's posture and are complete with features such as tails, horns and antlers. Clearly the Palaeolithic artists were capable of depicting realistic life forms when the context required it. By analogy with the Chinchorro mummies, the images of 'people' in Palaeolithic art may not relate to members of the artist's community, but may have represented inhabitants of a spiritual world.

The most famous of the Palaeolithic anthropomorphic representations are the 'Venus figurines', stylized portrayals of females which were made from a range of raw materials including clay, soft stone and ivory. A few of these figurines have exaggerated proportions with large breasts and protruding stomach and buttocks, but there are actually a wide range of body types portrayed. The Venus figurine found at Laugerie-Basse in 1864, the first discovered in modern times, has a relatively slim body even by the standards of contemporary fashion magazines! The figurines usually lack any details of the head or facial features – the few examples of realistically carved ivory heads that have been found are of doubtful

79. Venus figurine from Lespugue, France. In contrast to realistic and life-like portrayals of animals, ice age representations of the human form were often highly stylized and details of the head and face are often lacking.

80. Ivory statuette from Hohlenstein-Stadel, Germany, radiocarbon dated to about 30,000 years ago. The figure combines elements of human and animal form.

provenance and may even be modern forgeries carved into fossil ivory.

One early ivory figurine, a feline-headed statuette from the cave of Hohlenstein-Stadel near Ulm in Germany, combines elements of human and animal form and has been radiocarbon dated to the early Upper Palaeolithic, about 30,000 years ago.[23] The figurine was recovered in many fragments during an archaeological excavation in 1939, but the find was only recognized several decades later when the carved fragments of ivory

81. *Painted and engraved figure of the 'Sorcerer' in Trois Frères Cave, France. Such figures have been interpreted as depicting ritual specialists who dressed as animals, but they may instead represent beings inhabiting imaginary worlds.*

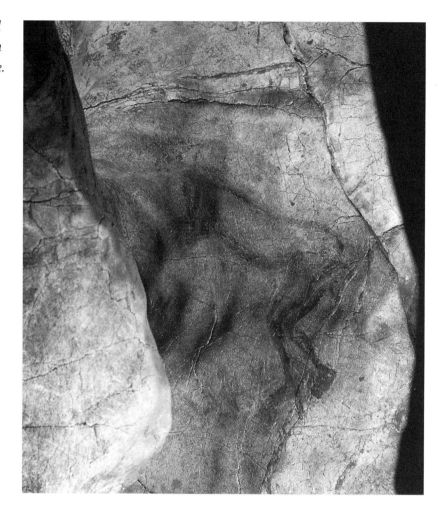

were discovered in a box of animal bones from the excavation. The figure stands upright and has human-proportioned legs but the head, and perhaps the forelimbs, resemble those of a carnivore, possibly a bear or a lion. Other composite figures that combine elements of human and animal form are two engraved figures from the Trois Frères cave in France, one depicting a bison's head on a human body, and the other a human torso and limbs but with a tail and the head of a stag. These have been interpreted as shamans – ritual specialists dressed as animals – but equally they may not be representations of living people at all, and might best be viewed as beings inhabiting a domain that was separate from that of the living.

CHAPTER 6

To infinity and beyond?

THE EMBALMING OF CORPSES IN CONTEMPORARY BRITISH AND AMERICAN CULTURE

> *... my executor will from time to time cause to be conveyed to the room in which [my friends and disciples] meet the said box or case with the contents therein to be stationed ...*
>
> (Excerpt from Jeremy Bentham's will of 30 May 1832, referring to the attendance of his embalmed corpse at council meetings at University College London)

One of the strangest exhibits on public display in London is the wax effigy of Jeremy Bentham, the eighteenth- to nineteenth-century philanthropist and utilitarianist philosopher. When Bentham died in 1832, he left instructions for his body to be preserved and exhibited after it had been publicly dissected. It was to be carried every so often into the university council chamber so that it might attend committee meetings. Bentham's body can still be seen in a corridor of University College London, but today the waxwork which sits in a glass case is no longer truly a body. The skeleton is still there, hidden beneath the wax, but the flesh has gone. The preserved head still survives but this is now kept locked away; the head sitting atop the wax body is itself of wax. However odd this preservation of a body may seem, Bentham had a serious aim. He wanted to broadcast a pragmatic and humanist approach to death and the disposal of the body, to try and counter, in his view, the irrational superstitions and beliefs fostered by religion. Bentham's wishes have perhaps been fulfilled with religion on the decline in Britain since his time.

Although Bentham's effigy is an 'auto-icon' rather than a mummy, when he died the skills and techniques were already

available to perform elaborate chemical mummification. Dr William Hunter of the Royal College of Surgeons was one of those particularly interested in preserving anatomical specimens, as aids to the understanding of the workings of the human body. In 1775 he and his brother John embalmed the body of Maria Van Butchell. They injected turpentine and vermillion into the cadaver's arteries, removed the thoracic and abdominal organs and washed the whole body in spirits of wine. After injecting the body for a second time, they returned the viscera (which had been treated with camphor, nitre and resin) and finally sewed up the body, filling the orifices with camphor and rubbing it all over with oils.[1] Maria's husband, Martin Van Butchell, was a well-known London eccentric and allowed the public to visit his mummified wife. After Van Butchell's death, his son presented his mother's mummy to the museum of the Royal College of Surgeons where she remained on display until destroyed during the blitz of The Second World War. At about the same time, the French anatomist Honoré Fragonard was conserving dissected human cadavers by injecting them with chemicals.[2]

For nearly 200 years thereafter, chemical embalming was a method associated in Britain only with the preservation of bodies donated to science. It was performed very rarely on any other dead bodies except those requiring long-distance shipment. In the last fifty years British readers were startled and shocked by two books on American funeral rites. One was Evelyn Waugh's *The Loved One*,[3] a novel about a pet mortician in California, and the other was Jessica Mitford's *The American Way of Death*, a journalistic exposé of the commercialized practices of American undertakers.[4] For British audiences the American obsession with embalming was then considered most peculiar. And yet it is commonplace in Britain today, euphemistically described by undertakers as 'hygienic treatment' or 'temporary preservation'.

Modern embalming involves the draining of the blood and the injection of preserving fluid into the arteries. Injection can be performed with syringe, hand pump or electric pump, gently

82. Wax effigy of the utilitarianist philosopher Jeremy Bentham, who wished his body to be exhibited after his death.

forcing the blood through the circulatory system and out of an opened vein. After the body is washed, the facial features are set in a natural position by closing the mouth up with needle and thread, massaging cream into the lips to keep them moist and closed, and preventing the eyeballs from sinking into their sockets either by cotton-wool padding or injection of a cream. Embalming fluids normally contain a weak solution of less than 2% formaldehyde (formalin when dissolved in water), together with alcohol, glycerine, borax, phenol, potassium and colouring dyes. A single corpse might require as much as two gallons, estimated by the body's weight in stones – one pint for one stone – and adding an extra pint.[5] This method prevents the onset of putrefaction for eight weeks or more; to embalm a body for longer requires injection through not one but six arteries, followed by spraying of the corpse with amyl acetate to prevent the growth of surface mould.[6]

Nineteenth-century American morticians developed this arterial method of draining the corpse of its blood and substituting a solution of chemicals for which the basic ingredients remain much the same today. This extraordinary intervention with the corpse satisfied two central aspects of American culture and funerary rites. One was hygiene and the other was the requirement that the dressed corpse should be displayed within an open coffin or casket. Late nineteenth-century society on both sides of the Atlantic had become increasingly fixated on hygiene and sanitation in every aspect of life, and dead bodies were no exception, posing potential health risks if not disinfected.[7] In Britain this concern seems to have been addressed initially by the use of spacious suburban burial grounds to relieve the overcrowded urban churchyards and then, during the twentieth century, by the construction of crematoria.

In America, where cremation never caught on, the emphasis has remained on the need for display. Coffins were no longer simple and cheap wooden boxes but handsome and elaborate caskets, made of expensive woods or metals and furnished to provide maximum 'comfort' for the corpse. Viewing of the body has been a necessary social element of American funerary rites. Funeral parlours provide whole wardrobes of burial outfits for the dead, from specially designed shoes to suits, dresses and negligees. Careful attention is paid to the face of the corpse, applying tints and make-up, and moulding the facial features so that the deceased has the appearance of life. In the land of the automobile, drive-by funeral parlours have made possible the paying of last respects to the deceased without even getting out of the car.

The American way of death contains within in it strange contradictions. Presentation of the corpse to family and friends is the clearest and most vivid confirmation that a person is indeed dead. Yet the metaphorical associations of sleeping in a cosy casket and the euphemisms of undertakers' discourse – dearly departed, gone to sleep, resting in the funeral parlour's slumber room, 'interred' rather than buried – all serve to present a negation or denial of death.[8] By the time of burial the body of a 'loved one' has been interfered with externally and internally, by persons not formerly known to the deceased, to a degree that is on a par with mummification in ancient Egypt.

In Britain, embalming has been steadily gaining ground since the late 1970s. At that time it was a practice mainly employed for treating corpses that had to travel time-consuming distances or, owing to family or political circumstances, required a delayed burial. Bodies donated to science were also embalmed in order to preserve them for later dissection by medical students. Only a few communities, such as travelling fairground showmen and Gypsies, required the elaborate and lengthy viewing of the body for which embalming was recommended. A handful of undertakers were beginning to take courses in embalming and to construct the blood-draining facilities within their mortuaries. But during the next twenty years the British undertaking trade was transformed.

American funeral businesses attacked the British market, buying up or driving out many of the small family concerns which had been the bedrock of the industry. Companies such as SCI (Service Corporation International) introduced an

aggressive style of business, marketing expensive 'hygienic' practices that included embalming and a widening choice of elaborate funerary fittings and services.[9] Today, with the agreement of the bereaved, embalming is carried out in most cases. In addition, one in five corpses will have been subject to medical post-mortems, to establish the cause of death, in which the skull will have been sawn in two, the chest area opened up and the internal organs explored. Many generations ago the British public handed over responsibility for their dead to professionals unrelated to the family and seem not to have noticed the revolution which has now occurred in British funerary practices. In a very short period of time, Britain has joined America in becoming a 'preserved body' culture, successors to the ancient Egyptians in the preservative treatment of the corpse, though with undoubtedly different motivations.

In archaeological terms it is hard to say how this will manifest itself for archaeologists of the future. Only 30% of corpses in Britain are buried, but these, and the majority of American burials, will rot considerably more slowly as a result of having been embalmed. Lead-lined caskets are also in vogue towards the top of the range and their potential for long-term preservation can also be remarkable, as archaeologists have discovered when excavating medieval and eighteenth- to nineteenth-century examples.[10] A growing interest in 'green' burial – being buried in biodegradable coffins with a tree planted on top – reflects our concerns for the environment and for our remains to be recycled as nutrients within the food chain. In Britain, the euphemism of sleep sits uneasily in our minds alongside the idea that dead bodies must be incinerated as unwanted waste. In addition to the large amount of gas required to fuel cremation ovens, each year in Britain about a million gallons of toxic embalming fluid are either burned in cremations or enter the groundwater as a potential pollutant from burials.

So shrouded in denial and silence has been the Western way of death that most of us have little idea just how bizarre and extraordinary is our own treatment of the dead. We are

83. Gateway in the Egyptian style, Sheffield General Cemetery. This imposing entrance emphasizes the permanence of death and the separation of the cemetery from the world of the living.

84. Entrance to a 'green' cemetery. In contrast to the entrance shown in fig. 83, this unobtrusive gateway, constructed in a familar rural style, provides a peaceful and relaxed setting that welcomes the visitor.

fascinated by the exotic strangeness of other cultures, past and present, who preserve the dead in embalmed or mummified form and yet we cannot even see that our own practices are really little different. For those who protest that our motivations are grounded in practical concerns with hygiene rather than in some arcane and peculiar religious beliefs about the afterworld, we need to understand that 'hygiene' is itself a way of thinking, a system of belief as much as a practical necessity. Our dead are, in any case, kept refrigerated prior to despatch – embalming is, in most circumstances, not necessary at all for public health reasons. As Jessica Mitford revealed in her exposé of the American way of death, embalming serves no useful purpose in preventing the communication of disease and is simply part of the commercial marketing of a more expensive funeral package.[11] This is not to say that embalming in contemporary society is something that is 'wrong'. Rather, it should be recognized as a cultural style, as a way of doing things which relates to people's often unformed notions about keeping the corpse intact for as long as possible in denying that 'the end' has come, and about the treatments necessary to render it unthreatening, unpolluting and harmless in ways that are as much spiritual as microbiological.

THE SEARCH FOR IMMORTALITY

> *I don't want to achieve immortality through my work . . .*
> *I want to achieve it through not dying!*
> (Woody Allen)

In January 1967 Professor James Bedford of California, USA, became the first person to have his body frozen with the intention of being restored to life at an unspecified future date.[12] James Bedford was an adherent of cryonics, the practice of preserving human corpses by deep freezing them in order to prevent the natural process of decomposition. The technology involved in cryonic preservation may be modern, but the quest for immortality is an ancient theme that appears in the earliest recorded myths.[13] The use of cryogenic (low-temperature) techniques for arresting the natural processes of decay is, however, a recent practice that became popularized in the USA in the 1960s following the publication of *The Prospect of Immortality* by Robert Ettinger.[14] Ettinger brilliantly combined science fact – the proven use of cryogenic procedures to temporarily freeze living cells – with the optimistic and some would say fanciful notion that future scientific advances would make it possible to resuscitate whole organisms from a deep-frozen state.

Some proponents of cryonic preservation refer to the frozen body as being in a 'suspended' state of life, but in reality it is illegal to freeze a body before clinical death has taken place, and in the eyes of the law a frozen body is just as dead as one which has been buried in the ground or cremated. The procedures of cryonic preservation are intended firstly to prevent the destruction of cells and tissues which commences at death, and secondly to minimize any further damage that freezing inevitably causes to the delicate cellular structures of the human body. As soon as death has been confirmed the body's circulatory system is connected to a machine which cools the blood rapidly and allows the body to be perfused with chemicals that reduce tissue damage. The body is then further cooled in a bath of silicone oil to a temperature of -80°C. Finally it is wrapped in thermal insulation inside a sealed aluminium container before being transferred to a liquid nitrogen container where the body is cooled to its final resting temperature of -196°C.[15]

Cryonic preservation is undertaken by private organizations and the expenses, which are considerable, are usually covered by investments or insurance policies arranged by the subject prior to his or her death. The practice does not enjoy the support of established medical scientists, who are sceptical about the future prospects of reversing the conditions that resulted in the individual's death let alone the problem of correcting the additional damage to the cells of the body that results from the preservation process itself. Supporters of cryonic preservation point to the fact that some animals can undergo short-term freezing without apparently suffering ill effects. Some fish can survive several months of freezing within the ice of freshwater lakes, and an experiment has been reported in which frost-tolerant frogs survived being exposed to a temperature of -6°C for five days, by which time one-third of their body fluids had frozen.[16] However, these animals were adapted to winter conditions and high levels of the anti-freezing agent glycerol in their blood prevented their bodies from freezing completely. Humans too can survive brief episodes of hypothermia, with cases of survival and recovery being reported after prolonged cooling of the body in icy water. However, a frozen human body has never been revived successfully.

Critics of cryonic preservation point out that while individual cells can be frozen and then revived under laboratory conditions (for example, the freezing of sperm for later use in artificial insemination), the specific requirements of the many different tissues of the body cannot be satisfied by a single protocol of cooling and freezing: different cells require different conditions of freezing and thawing. Furthermore, it is extremely doubtful that adequate function could be restored to an organ as complex as the human brain: the brain cells are damaged by just a few minutes of oxygen deprivation at the temperature of the living body, and the body cannot easily be cooled fast enough to prevent this damage from occurring. Cryonics organizations also discourage any procedures such as autopsy and organ donation which damage the cadaver, and this is a further point of potential conflict with the medical profession.[17]

In the 1960s the pioneers of cryonic preservation believed that future advances in miniaturized technology (so-called 'nanotechnology') would provide the means of repairing and restoring to life the damaged and worn-out tissues of the frozen subject. But advances in science are almost impossible to predict, and in the few decades that have passed since the body of James Bedford was lowered into its chilly tomb there have been unforeseen developments in molecular biology and in computing that

85. *Dolly the cloned sheep. Cloning provides the possibility of perpetuating the life of an individual organism, but its application to humans raises many legal and ethical questions.*

have offered glimpses of alternative routes to immortality. Animals have been cloned from a single cell,[18] and although it is not legally permitted to apply this method to humans it suggests an alternative and more plausible means of sustaining the life of the individual for an indefinite period of time.

Advances in the power and speed of computers have enabled the development of artificial intelligence and virtual reality, the realistic modelling of the world through computer memory and video graphics. Electronic replicas of people (so-called 'avatars') already answer many telephone enquiries and present information on websites and television. Insofar as it may become possible to replicate the appearance and personality of an individual in computer memory, the frail and fallible 'wetware' of the human body might conceivably become redundant!

The Ethics of Display and Ownership

We owe respect to the living: to the dead we owe only the truth.
(Voltaire, *Première lettre sur Oedipe*)

Most people like to think that they treat corpses with dignity and respect, that the dead are allowed to 'rest in peace'. Anyone who has witnessed the commercial clearance of a graveyard knows that this is not the case. Coffins, corpses and skeletons are bull-dozed and bucketed out of their graves in ways that are anything but respectful in order to make way for sale of church land, retail development, and extensions or repairs to church buildings. The work is carried out by commercial undertakers who, naturally enough, do not publicize their activities. In a growing number of cases where eighteenth- and nineteenth-century cemeteries and graveyards come under threat, archaeologists may carry out the work of excavating and recording. Even though their work is careful, respectful and meticulous, and will result in the reburial of the remains, archaeologists often come under fire from misinformed members of the public who object to the dead being disturbed. Of course, archaeologists' work is normally in the public domain and is often subject to closer scrutiny than a commercial clearance undertaker. As a result, those who com-plain about archaeological cemetery excavations probably have little idea about the planning process which has led to the burials having to be disturbed. They are also unlikely to be aware of the more cursory and brutal methods of commercial undertakers who may be motivated by profit rather than by the search for knowledge which underlies archaeologists' interests in the dead.

Few people in Britain come into direct contact with corpses on a regular basis. Nurses, doctors, pathologists, hospital staff, police, firemen, ambulance workers, undertakers and occasion-ally archaeologists are amongst those who do. We may not even see the dead bodies of our nearest and dearest if they die in hos-pital. Perhaps part of the reason why corpses are so fascinating is precisely because we see them so rarely. This morbid curiosity may be born out of their absence from 'real life', however many we may see in films or on the television. Perhaps also, like the slave at the Roman emperor's shoulder, they serve to remind us that we too are mortal and must surely die. Our culture has been described by philosophers and sociologists as one in denial of death. By sweeping the remains of the dead under the carpet we are likely to foster an approach to life which attempts to shut out death and which refuses to confront the process of dying.

There is a tiny proportion of the population in Britain who have misgivings about the display of ancient corpses. Every once in a while someone writes to protest about Lindow Man being on display in The British Museum. But for every one person that protests there are many thousands who join the cluster of people gathered around to look at his peat-stained body. For most people there is nothing more amazing about the past than looking into the face of someone who died hundreds or thousands of years ago. Here is the heart of the mystery – this person lived, loved and died and we can study the lines on their face, the cut of their hair, and the clothes they wore, but what were they actually like? And how and why did they die? And what was life actually like for them and others in their community? The past is at once so close and so real and yet we realize how very far away it is and how we can never ever really grasp what it was like.

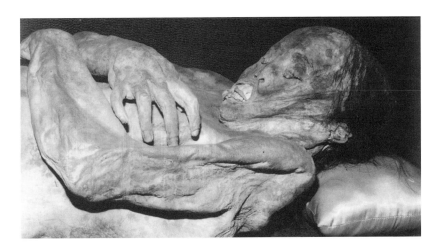

86. Naturally mummified body on display in the Museo de las Momias, Guanajuato, Mexico. Attitudes to the display of the dead vary amongst present-day cultures, and are subject to changing cultural norms.

In a small museum in Guanajuato, Mexico there is an extraordinary and, to some, shocking display of dried corpses. These had been buried in the underground catacombs about a hundred years ago and were rediscovered by archaeologists. Until recently, the bodies, many of them clothed, were displayed so that there was nothing between them and the museum visitor. After a number of thefts and acts of minor vandalism they are now behind glass but they still retain their extraordinary and disturbing power.[19]

Mexico is famous for its Day of the Dead and accompanying relaxed attitudes to death and the dead. Yet there are other communities in North America which are far from relaxed, and generally with good reason. The astute reader will have noticed that this book contains no pictures of indigenous mummies from North America. Over the past thirty years Native American communities have striven to reclaim their dead ancestors from

87. Sugar skulls from the Mexican Day of the Dead. Relaxed attitudes to death and the dead are a characteristic of Mexican culture.

88. *Reburial of the skeletons of Native Americans from Wounded Knee. Reburial is one manifestation of the regaining of control by indigenous peoples over their cultural and physical heritage.*

the clutches of museums and private collectors, and to rebury all human remains from cremated bones to fully preserved bodies. They do so with the full support of the law – the Native American Graves Protection and Repatriation Act 1990 (NAGPRA) – and with five centuries behind them of desecration, removal and stealing of their dead by Europeans for curios, museum exhibits and scientific specimens. For people who have been removed from their ancestral lands onto reservations, discriminated against by incoming groups and treated like second-class citizens in their own land, and whose cultures have reached the verge of extinction within the totalizing American way of life, reclaiming control over the remains of the dead is one arena where they can reassert that pride and dignity. Many North American archaeologists – not of indigenous descent – have not been at all happy about this political manoeuvring because their study material has been taken away. Archaeologists have quite naturally resented being told that the information that they can learn and communicate about ancient Native American populations is irrelevant because Native Americans know all they need to from myth and tradition. Today the situation has calmed down in many states because archaeologists and Native Americans have begun to treat each side's considerations with more respect.

It is not just native North Americans who may hold antagonistic attitudes about displaying bodies. Fundamentalist Christians may also have religious objections and, in addition, may not be kindly disposed towards accompanying views of the world which dismiss their creationism. In a small visitor survey carried out by English Heritage in its local museum at Avebury, most people approved of keeping the Neolithic child skeleton on display, but a surprising number of those against its continued display were American tourists.

The issues surrounding the display, keeping and reburial of human corpses are very complex. On one level, certain people are outraged or disgusted by the 'pornography of death', by the shocking and graphic image of a dead human. On another, people may feel that we, the living, are treating the dead disrespectfully by displaying, storing or otherwise interfering with their corpses in ways which might have been against their wishes or against their cultural traditions. Those who consider themselves to be the living descendants, who claim those dead as ancestors, may also consider that their own dignity and self-respect have been violated and affronted.

Perhaps at the root of these powerful feelings are relationships of inequality between the powerful and the powerless, and challenges to emotional security. Colonized indigenous groups throughout the world have suffered from exploitation, genocide, starvation and disease. Alcohol and Western lifestyles have eroded cultural traditions and European incomers have continued to insult and abuse. Reactions have been strongest where the Europeans have maintained their colonial domination, in the United States and Canada, in Australia and New Zealand, and in South Africa. Interestingly, there has been little reaction in the Catholic nations of Central and South America where mummies and skeletons are either big business for tourism or sources of illegal income for looters.

Australian Aboriginals have good cause to have been concerned about the predatory despoilations of European scientists who turned the remains of their ancestors into trophies and specimens. In 1907 the German scientist Hermann Klaatsch obtained a newly dead Aboriginal corpse and shipped it to Germany for dissection. In the 1920s Sir John Cleland, professor of pathology at the South Australian Museum, placed four Aboriginal corpses in tanks of preserving fluid so that they could be retained for study as specimens of Aboriginal 'pure-bloodedness' – 150 years of white men raping and sexually abusing black women had contributed to the shortage of 'pure-blooded' specimens. In 1869 the corpse of the last Tasmanian male of unmixed descent, William Lanney, was dissected. Subsequently William Crowther of London's Royal College of Surgeons took his skull and allegedly the preserved skin of his nose, ears and face, making a portion of his skin into a tobacco pouch. The reason for this intrusive level of interest by scientists was that Australian Aboriginals were considered to be evolutionary throwbacks – either relatives of the Neanderthals or missing links between apes and humans. Aboriginals were, in one sense, victims of the Western Enlightenment idea of progress in which Victorian social evolutionary models had placed them at the lowest level of 'savagery'.

Conciliatory attitudes and protective legislation developed across Australasia in the 1980s and 1990s. Amongst the thousands of skeletons that have been returned for reburial or for Aboriginal keeping are numerous mummies. One form of mummy which is still highly valued around the world is the Maori tattooed head. These were formerly collected by the Maori themselves, as headhunted trophies or as venerated ancestors. In the years after Captain Cook's voyage, these heads were sold, given to and stolen by Europeans who bought and sold them as collectibles for private collections and for public museums. Up until the 1990s tattooed Maori heads might be auctioned for thousands of pounds in the major auction houses of the world. It was only recently that Sotheby's withdrew a head for sale prior to auction as a result of a protest by Maori representatives. Even those heads given as gifts by Maoris to Europeans have been requested for return.

The display of non-European corpses is broadly unacceptable throughout much of the world today. In 1992 an exhibited preserved corpse of a Southern African Khoi !San (formerly called Bushmen) in a museum in Barcelona became the focus of an Olympic Games protest when delegates attending Olympic events in Barcelona were outraged to find this legacy of colonialist chauvinism.

Today there are international codes of ethics to protect indigenous peoples against the thoughtless appropriation, curation and display of their ancestors' remains. The International Council on Museums (ICOM) specifies that '[a]lthough it is occasionally necessary to use human remains and other sensitive material in interpretive exhibits, this must be done with tact and with respect for the feelings of human dignity held by all people'. In 1989 the World Archaeological Congress, which has done much to change archaeologists' attitudes to indigenous human remains, issued a statement on archaeological ethics and the treatment of the dead, known as the Vermillion Accord. It stresses the need for mutual respect for the beliefs of indigenous peoples as well as the importance of science and education, and takes into account the wishes of the dead where known or inferred and the wishes of the local community.

We have lived in a globalized society since at least 1700 and today the degree of knowledge transfer and communication, especially through television and the internet, is on a scale unprecedented. Sociologists and anthropologists have discussed over and over whether this globalization will lead to a homogenized 'jeans and cola' world culture in which local traditions and communities are swamped. The answer at present seems to be that the opposite is happening. Local cultural and ethnic identities have revitalized and flourished in the face of multinational and federal encroachment. A tiny part of this change has centred around the politics of display and reburial. So many local and indigenous groups have demanded reburial and restitution that this attitude is almost becoming the new global orthodoxy.

So should Lindow Man and other European bog bodies be reburied? Our answer is an unequivocal 'no'. Whereas non-disturbance may be a sacred tradition to be respected in other cultures, it certainly is not in ours. Whether we look to the past – ancient or recent – or to the present, people in Britain and Europe have been exhuming, swapping, cutting up, stealing and

89. *Skull from an excavation of a nineteenth-century hospital burial ground. The top of the skull had been removed during an autopsy that took place prior to burial. Scientific knowledge was gained from this body in the past, just as archaeologists today learn from excavations of human remains.*

exhibiting corpses or parts of them. 'Resting in peace' is not an accurate description of the dead and how they have been treated over several millennia. Every society lives intimately with its dead in one form or another. In addition, one aspect of European cultural tradition is that the dead can also contribute through their actual physical remains to knowledge, health and inspiration about the meaning of life, the universe and everything.[20] Preserved bodies help us to understand our origins and our past, in both biological, and social and cultural terms. They provide a unique insight into ancient lives, creating direct connections with the past that ordinary artefacts can never make. We learn from post-mortems of the ancients as well as the newly dead.

Preserved bodies have had an enormous impact on all those who have seen them. Owen Beattie has described the staggering worldwide public reaction to photographs of the corpses of the two sailors who were buried on the ill-fated Franklin expedition of 1845–8.[21] In particular, that disturbing and eerie face of the young John Torrington has inspired playwrights, poets, painters, sculptors, photographers and even the heavy metal band Iron Maiden.

In the words of the poet Seamus Heaney, when we examine the bodies of the deceased, we come 'face to face with our past'. They have 'a double force, a riddling power: on the one hand, they invite us to reverie and daydream, while on the other hand, they can tempt the intellect to its most strenuous exertions. And it has always seemed to me that this phenomenal potency derives from the fact that the bodies erase the boundary-line between culture and nature, between art and life, between vision and eyesight.'[22]

Notes

Chapter 1

1 Balguy 1734: 413
2 Krogman and Iscan 1986
3 Payne 1965
4 Rodriguez and Bass 1983, 1985; Mann *et al.* 1990
5 Bell *et al.* 1996
6 Mant 1987; Janaway 1996
7 Ascenzi *et al.* 1998
8 Danforth 1982
9 Hallam 1982: 359
10 Horrox 1999: 99; Daniell 1997: 44; Litten 1991: 35–43
11 Ayloffe 1775
12 Kantorowicz 1957; Metcalf and Huntington 1991: 165–72; Binski 1996: 61
13 Litten 1991: 160
14 Fraser 1973
15 Cobo [1653] 1990: 127–8; *see also* Isbell 1997: 38–68 on royal Inca mummies
16 Sillar 1992
17 Sillar 1992: 112
18 Cited in Isbell 1997: 38
19 Neely 1982: 121–2; Power 1872
20 Neely 1982: 309–10
21 Stalin quoted in Tumarkin 1983: 174
22 Tumarkin 1983
23 Smart 1998: 531
24 Frayling 1992
25 Tumarkin 1983: 180
26 Remnick 1994: 147
27 Remnick 1994: 220–21
28 Remnick 1994: 504
29 Hyde 1971: 596–8
30 Radzinsky 1996: 560
31 Deutscher 1967: 630
32 Terrill 1980: 422
33 Alexander 1979: 89
34 Alexander 1979: 88–91
35 Doumerc 1989: 119; Dujovne Ortiz 1997: 344
36 Dujovne Ortiz 1997: 346
37 Main 1977: 277–80
38 Main 1977: 285

Chapter 2

1 Vreeland 1978; 1998
2 Doran *et al.* 1986; Hauswirth *et al.* 1991
3 Royal and Clark 1960
4 Oakley 1960; Tkocz *et al.* 1979
5 Painter 1995
6 Ó Floinn 1995; van der Sanden 1995
7 Turner 1995, 1999
8 Garland 1995
9 Bennike 1999
10 Skaarup 1985
11 Bennike and Ebbesen 1987
12 Bennike *et al.* 1986; Bennike 1985
13 Bennike and Bro-Rasmussen 1989; Bennike 1999
14 Lillie 1999
15 Margetts 1967
16 Becker 1947; Bennike and Ebbesen 1987; Thorpe 1996: 138–9; Koch 1998, 1999
17 Koch 1998, 1999
18 Andersen and Geertinger 1984
19 Glob 1969
20 Munksgaard 1984
21 Parker Pearson 1984
22 Andersen and Geertinger 1984
23 van der Sanden 1996: 136
24 Fischer 1999
25 Gebühr 1979
26 Stead *et al.* 1986
27 Stead *et al.* 1986
28 Brothwell 1986
29 Ross and Robins 1989
30 Parker Pearson 1986; Magilton 1995
31 Pyatt *et al.* 1991; Pyatt and Storey 1995
32 May 1996
33 Connolly 1985
34 Field and Parker Pearson forthcoming
35 Bradley 1990; Field and Parker Pearson forthcoming
36 Bennike 1993; 1999: fig. 6.3
37 Magilton 1995
38 Fischer 1998
39 Hedeager 1990; Parker Pearson 1984
40 Cited in van der Sanden 1996: 167
41 van der Sanden 1996: 167
42 Munksgaard 1984
43 van der Sanden 1996: 177
44 Bloch Jørgensen *et al.* 1999
45 Munksgaard 1984
46 Fischer 1998, 1999
47 van der Sanden 1996: 180

Chapter 3

1 Vreeland 1998

2 Harkin 1990

3 Pretty and Calder 1998

4 Arriaza *et al.* 1998

5 Vreeland 1998

6 Tapp 1985; Brothwell 1986

7 Arriaza 1995a–c; Arriaza *et al.* 1998; Schreiber 1996

8 Reid 1999: 205–6

9 Spencer 1982; Partridge 1994; David 2000; Taylor 2001

10 Partridge 1994; Andrews 1984; Wallis Budge 1987

11 Taylor 2001: 47–8, 79

12 Emery 1961: 162–4; Adams 1984: 9–10; Spencer 1982: 35

13 Lehner 1997: 88

14 Partridge 1994: 6

15 Partridge 1994; Taylor 2001

16 Taylor 2001

17 Lucas 1932, 1962; Snape 1996

18 Brier 1996; Brier and Wade 1997

19 Carter 1927

20 Taylor 2001: 84

21 Pretty 1969

22 Balabanova *et al.* 1992

23 Child and Pollard 1992

24 Springfield *et al.* 1993

25 Panagiotakopulu and Buckland 2001

26 Barley 1995: 112

27 Adams 1977; Huntington and Metcalf 1979

28 Spencer 1982: 128–9

29 Litten 1991: 37; Barley 1995: 106

30 For example, Barley 1986: 103

31 Huntington 1987; Mack 1986

32 Harrison *et al.* 1969

33 Pääbo 1985a, 1985b

34 Hughes *et al.* 1986; Reyman *et al.* 1998

35 Handt *et al.* 1994

36 Rollo *et al.* 2000

37 Smith *et al.* 2001

38 Barber 1999; Mallory and Mair 2000

39 Francalacci 1995

Chapter 4

1 Gamble 1993

2 Beattie and Geiger 1987

3 Kowal *et al.* 1990

4 Kowal *et al.* 1991

5 Farrer 1993

6 Keenleyside *et al.* 1996

7 Hansen *et al.* 1991

8 Lobdell and Dekin 1984; Dekin 1987

9 Zimmerman and Aufderheide 1984

10 Zimmerman and Smith 1975

11 Smith and Zimmerman 1975

12 Kuehn 1999

13 Spindler 1994

14 Seidler *et al.* 1992; Spindler 1994

15 Rollo *et al.* 2000

16 Horne *et al.* 1982; Vreeland 1998; McEwan and Van de Guchte 1992

17 Reinhard 1996, 1997

18 Vreeland 1998

19 McEwan and Van de Guchte 1992

20 Rudenko 1970

21 Polosmak 1994; Bogucki 1996

22 Parker Pearson 2000

Chapter 5

1 Carver 1998: 66

2 Carver 1998

3 For example, Ashbee 1957

4 Lindsay 1960; Gore 1984; Descoeudres 1994; Gill 1996

5 Grant 1971: 34–8

6 Allison 1994

7 Leakey and Hay 1979; Leakey 1986

8 Robbins 1986

9 Aldhouse-Green *et al.* 1992

10 Roberts *et al.* 1996

11 Bahn and Vertut 1997

12 Faulds 1880

13 Mania and Toepfer 1973

14 Åström 1980

15 Kamp *et al.* 1999

16 Fletcher 1979

17 Clottes *et al.* 1992, 1997

18 Layton 1992

19 Layton 1992; Flood 1997: 225

20 Janssens 1957; Barrière 1976; Bahn and Vertut 1997

21 Moore 1977; Walsh 1979

22 Bahn and Vertut 1997: 218

23 Hahn 1971

Chapter 6

1 Litten 1991: 46–50

2 Ellenberger 1981

3 Waugh 1948

4 Mitford 1965, 1998

5 Polson and Marshall 1975: 339–42

6 Polson and Marshall 1975: 351–2

7 Chidester 1990: 278; Parker Pearson 1982

8 Becker 1973; Huntington and Metcalf 1979; Chidester 1990: 278–80

9 Jupp and Walter 1999

10 O'Sullivan 1982; Reeve and Adams 1993; Cox 1996, 1998; Reeve and Cox 1999

11 Mitford 1965, 1998

12 Nelson 1968

13 Parker Pearson 2000: 168

14 Ettinger 1964; *see also* Sheshkin 1979

15 Iserson 1994: 293

16 Schmid 1982

17 Iserson 1994: 292

18 Wilmut *et al.* 1997

19 Medina Monzón 1993

20 See, for example, Finn 1999

21 Beattie 1999

22 Heaney 1999: 4

Bibliography

Adams, B. 1984. *Egyptian Mummies*. Princes Risborough, Shire.

Adams, M.J. 1977. Style in southeast Asian materials processing: some implications for ritual and art. In H. Lechtman and R. Merrill (eds), *Material Culture: studies, organisation, and dynamics of technology*. St Paul, West Publishing, pp. 21–52.

Aldhouse-Green, S.H.R., Whittle, A.W.R., Allen, J.R.L. *et al.* 1992. Prehistoric human footprints from the Severn Estuary at Uskmouth and Magor Pill, Gwent, Wales. *Archaeologia Cambrensis* 141: 14–55.

Alexander, R.J. 1979. *Juan Domingo Perón: a history*. Boulder, Westview Press.

Allison, P. 1994. The Distribution of Pompeiian House Contents and its Significance. PhD Thesis, School of Archaeology, Classics and Ancient History, University of Sydney.

Andersen, S.R. and Geertinger, P. 1984. Bog bodies investigated in the light of forensic medicine. *Journal of Danish Archaeology* 3: 111–19.

Andrews, C. 1984. *Egyptian Mummies*. London, British Museum Press.

Arriaza, B.T. 1995a. *Beyond Death: the Chinchorro mummies of ancient Chile*. Washington DC, Smithsonian Institution.

Arriaza, B.T. 1995b. Chile's Chinchorro mummies. *National Geographic* 187 (3): 68–89.

Arriaza, B.T. 1995c. Chinchorro bioarchaeology: chronology and mummy seriation. *Latin American Antiquity* 6: 35–55.

Arriaza, B.T., Cárdenas-Arroyo, F., Kleiss, E. and Verano, J.W. 1998. South American mummies: culture and disease. In A. Cockburn, E. Cockburn and T.A. Reyman (eds), *Mummies, Disease and Ancient Cultures*. Second edition. Cambridge, Cambridge University Press, pp. 190–234.

Ascenzi, A., Bianco, P., Fornaciari, G. and Rodríguez-Martín, C. 1998. Mummies from Italy, North Africa and the Canary Islands. In A. Cockburn, E. Cockburn and T.A. Reyman (eds), *Mummies, Disease and Ancient Cultures*. Second edition. Cambridge, Cambridge University Press, pp. 263–88.

Ashbee, P. 1957. The great barrow at Bishop's Waltham, Hampshire. *Proceedings of the Prehistoric Society* 23: 137–66.

Åström, P. 1980. Fingerprints and archaeology. *Studies in Mediterranean Archaeology* 28. Göteborg.

Ayloffe, J. 1775. An account of the body of king Edward the first, as it appeared on opening his tomb in the year 1774. *Archaeologia* 3: 376–413.

Bahn, P.G. and Vertut, J. 1997. *Journey Through the Ice Age*. Berkeley, University of California Press.

Balabanova, S., Parsche, F. and Pirsig, W. 1992. First identification of drugs in Egyptian mummies. *Naturwissenschaften* 79: 358.

Balguy, C. 1734. An account of the dead bodies of a man and a woman, which were preserved 49 years in the Moors in Derbyshire. *Philosophical Transactions of the Royal Society of London* 38: 413–15.

Barber, E.W. 1999. *The Mummies of Ürümchi*. London, Norton.

Barley, N. 1986. *A Plague of Caterpillars: a return to the African bush*. Harmondsworth, Penguin.

Barley, N. 1995. *Dancing on the Grave: encounters with death*. London, John Murray.

Barrière, C. 1976. Palaeolithic art in the Grotte de Gargas. *British Archaeological Reports* Supplementary Series 14. Oxford, British Archaeological Reports.

Beattie, O. 1999. Sleep by the shores of those icy seas: death and resurrection in the last Franklin expedition. In J. Downes and T. Pollard (eds), *The Loved Body's Corruption: archaeological contributions to the study of human mortality*. Glasgow, Cruithne Press, pp. 52–68.

Beattie, O. and Geiger, J. 1987. *Frozen in Time. The Fate of the Franklin Expedition*. London, Collins.

Becker, C.J. 1947. Mosefundne lerkar. *Aarbøger for Nordisk Oldkyndighed og Historie 1947*.

Becker, E. 1973. *The Denial of Death*. London, Collier Macmillan.

Bell, M., Fowler, P.J. and Hillson, S.W. 1996. *The Experimental Earthwork Project, 1960–1992*. York, Council for British Archaeology.

Bennike, P. 1985. *Palaeopathology of Danish Skeletons: a comparative study of demography, disease and injury*. Copenhagen, Akademisk Forlag.

Bennike, P. 1993. The people. In S. Hvass and B. Storgaard (eds), *Digging into the Past*. Arhus, Arhus Universitetsforlag, pp. 34–9.

Bennike, P. 1999. The Early Neolithic Danish bog finds – a strange group of people! In B. Coles, J. Coles and M. Schou Jørgensen (eds), *Bog Bodies, Sacred Sites and Wetland Archaeology*. Exeter, WARP, University of Exeter, pp. 27–32.

Bennike, P. and Bro-Rasmussen, F. 1989. Contracture and ankylosis of the hip joint in a right angle: five palaeopathological cases. *International Journal of Anthropology* 4: 137–51.

Bennike, P. and Ebbesen, K. 1987. The bog find from Sigersdal. Human sacrifice in the Early Neolithic. *Journal of Danish Archaeology* 5: 85–115.

Bennike, P., Ebbesen, K. and Bender Jørgensen, L. 1986. Early Neolithic skeletons from Bolkilde bog, Denmark. *Antiquity* 60: 199–209.

Binski, P. 1996. *Medieval Death: ritual and representation*. London, British Museum Press.

Bloch Jørgensen, A., Robinson, D. and Christensen, C. 1999. Almosen, Denmark: a ritual bog site from the 1st millennium BC. In B. Coles, J. Coles and M. Schou Jørgensen (eds), *Bog Bodies, Sacred Sites and Wetland Archaeology*. Exeter, WARP, University of Exeter, pp. 121–3.

Bogucki, P. 1996. Pazyryk and the Ukok princess. In P. Bahn (ed.), *Tombs, Graves and Mummies*. London, Weidenfeld and Nicolson, pp. 146–51.

Bradley, R. 1990. *The Passage of Arms: an Archaeological Analysis of Prehistoric Hoards and Votive Deposits*. Cambridge, Cambridge University Press.

Brier, B. 1996. *Egyptian Mummies*. London, Michael O'Mara.

Brier, B and Wade, R.S. 1997. The use of natron in human mummification: a modern experiment. *Zeitschrift für Agyptische Sprache und Altertumskunde* 124: 89–100.

Brothwell, D. 1986. *The Bog Man and the Archaeology of People*. London, British Museum Press.

Carter, H. 1927. *The Tomb of Tut·Ankh·Amen*. II. London, Toronto, Melbourne & Sydney, Cassell & Co. Ltd.

Carver, M. 1998. *Sutton Hoo: Burial Ground of Kings?* London, British Museum Press.

Chidester, D. 1990. *Patterns of Transcendence: religion, death, and dying*. Belmont, Wadsworth.

Child, A.M. and Pollard, A.M. 1992. A review of the applications of immunochemistry to archaeological bone. *Journal of Archaeological Science* 19: 39–47.

Clottes, J., Beltrán, A., Courtin, J. and Cosquer, H. 1992. The Cosquer Cave on Cape Morgiou, Marseilles. *Antiquity* 66: 583–98.

Clottes, J., Courtin, J., Collina-Girard, J., Arnold, M. and Valladas, H. 1997. News from Cosquer Cave: climatic studies, recording, sampling, dates. *Antiquity* 71: 321–6.

Cobo, B. [1653] 1990. *Inca Religion and Customs.* Trans. and ed. R. Hamilton. Austin, University of Texas Press.

Connolly, R.C. 1985. Lindow Man: Britain's prehistoric bog body. *Anthropology Today* 1(5): 15–17.

Cox, M. 1996. *Life and Death in Spitalfields 1700 to 1850.* York, Council for British Archaeology.

Cox, M. (ed.) 1998. *Grave concerns: death and burial in England 1700–1850.* York, Council for British Archaeology.

Danforth, L.M. 1982. *The Death Rituals of Rural Greece.* Princeton, Princeton University Press.

Daniell, C. 1997. *Death and Burial in Medieval England: 1066–1550.* London, Routledge.

David, A.R. 2000. Mummification. In P.T. Nicholson and I. Shaw (eds), *Ancient Egyptian Materials and Technology.* Cambridge, Cambridge University Press, pp. 372–89.

Dekin, A.A. 1987. Sealed in time. *National Geographic* 171: 824–36.

Descoeudres, J.-P. 1994. *Pompeii Revisited: the life and death of a Roman town.* Sydney, Meditarch, University of Sydney.

Deutscher, I. 1967. *Stalin: a political biography.* London, Oxford University Press.

Doran, G.H., Dickel, D.N., Ballinger, W.E. *et al.* 1986. Anatomical, cellular and molecular analysis of 8000-yr-old human brain tissue from the Windover archaeological site. *Nature* 323: 803–6.

Doumerc, B. 1989. *Eva Perón.* Barcelona, Editorial Lumen.

Dujovne Ortiz, A. 1997. *Eva Perón: a biography.* London, Warner Books.

Ellenberger, M. 1981. *L'autre Fragonard.* Paris, Jupilles.

Emery, W.B. 1961. *Archaic Egypt.* Harmondsworth, Penguin.

Ettinger, R.C.W. 1964. *The Prospect of Immortality.* New York, Doubleday.

Farrer, K.T.H. 1993. Lead and the last Franklin expedition. *Journal of Archaeological Science* 20: 399–409.

Faulds, H. 1880. On the skin-furrows of the hand. *Nature* 28 October, p. 6.

Field, N. and Parker Pearson, M. forthcoming. *Fiskerton. An Iron Age Timber Causeway with Iron Age and Roman Votive Offerings.* London, English Heritage.

Finn, C. 1999. Words from kept bodies. The bog body as literary inspiration. In B. Coles, J. Coles and M. Schou Jørgensen (eds), *Bog Bodies, Sacred Sites and Wetland Archaeology.* Exeter, WARP, University of Exeter, pp. 79–83.

Fischer, C. 1998. Bog bodies of Denmark and northwestern Europe. In A. Cockburn, E. Cockburn and T.A. Reyman (eds), *Mummies, Disease and Ancient Cultures.* Second edition. Cambridge, Cambridge University Press, pp. 237–62.

Fischer, C. 1999. The Tollund Man and the Elling Woman and other bog bodies from central Jutland. In B. Coles, J. Coles and M. Schou Jørgensen (eds), *Bog Bodies, Sacred Sites and Wetland Archaeology.* Exeter, WARP, University of Exeter, pp. 93–7.

Fletcher, A. 1979. The fingerprint examination. In A.R. David (ed.), *Manchester Museum Mummy Project: Multi-Disciplinary Research on Ancient Egyptian Mummified Remains.* Manchester, Manchester University Press, pp. 79–82.

Flood, J. 1997. *Rock Art of the Dream Time.* Sydney, HarperCollins.

Francalacci, P. 1995. DNA analysis of ancient desiccated corpses from Xinjiang. *Journal of Indo-European Studies* 23: 385–98.

Fraser, A. 1973. *Cromwell the Lord Protector*. New York, Knopf.

Frayling, C. 1992. *The face of Tutankhamun*. London, Faber.

Gamble, C. 1993. *Timewalkers*. Stroud, Sutton.

Garland, A.N. 1995. Worsley Man, England. In R.C. Turner and R.G. Scaife (eds), *Bog Bodies: New Discoveries and New Perspectives*. London, British Museum Press, pp. 104–7.

Gebühr, M. 1979. Das Kindergrab von Windeby. Versuch einer 'Rehabilitation'. *Offa* 36: 75–107.

Gill, D. 1996. Herculaneum and Pompeii. In P. Bahn (ed.), *Tombs, Graves and Mummies*. London, Weidenfeld and Nicolson, pp. 154–9.

Glob, P.V. 1969. *The Bog People: Iron-Age man preserved*. London, Faber and Faber.

Gore, R. 1984. After 2,000 years of silence the dead do tell tales at Vesuvius. *National Geographic* 165: 556–613.

Grant, M. 1971. *Cities of Vesuvius: Pompeii and Herculaneum*. London, Weidenfeld and Nicolson.

Hahn, J. 1971. La statuette masculine de la Grotte du Hohlenstein-Stadel (Wurtemberg). *Anthropologie* 75: 233–44

Hallam, E. 1982. Royal burial and the cult of kingship in France and England, 1060–1330. *Journal of Medieval History* 8: 359–80.

Handt, O., Richards, M., Trommsdorf, M. *et al.* 1994. Molecular genetic analyses of the Tyrolean ice man. *Science* 264: 1775–8.

Hansen, J.P.H., Meldgaard, J. and Nordqvist, J. 1991. *The Greenland Mummies*. London, British Museum Press.

Harkin, M. 1990. Mortuary practices and the category of the person among the Heiltsuk. *Arctic Anthropology* 27: 87–108.

Harrison, R.G., Connolly, R.C. and Abdalla, A.

1969. Kinship of Smenkhkare and Tutankhamen demonstrated serologically. *Nature* 224: 325–6.

Hauswirth, W.W., Dickel, C.D., Doran, G.H., Laipis, P.J. and Dickel, D.N. 1991. 8000-year-old brain tissue from the Windover site: anatomical, cellular, and molecular analysis. In D.J. Ortner and A.C. Aufderheide (eds), *Human Paleopathology. Current Syntheses and Future Options*. Washington, Smithsonian Institution Press, pp. 60–72.

Heaney, S. 1999. The man and the bog. In B. Coles, J. Coles and M. Schou Jørgensen (eds), *Bog Bodies, Sacred Sites and Wetland Archaeology*. Exeter, WARP, University of Exeter, pp. 3–6.

Hedeager, L. 1990. *Iron-age Societies: from tribe to state in northern Europe, 500 BC to AD 700*. Oxford, Blackwell.

Horne, P.D., Kawasaki, S.Q. and Gryfe, A. 1982. The prince of El Plomo. *Paleopathology Newsletter* 40: 7–10.

Horrox, R. 1999. Purgatory, prayer and plague: 1150–1380. In P.C. Jupp and C. Gittings (eds), *Death in England: an Illustrated History*. Manchester, Manchester University Press, pp. 90–118.

Hughes, M.A., Jones, D.S. and Connolly, R.C. 1986. Body in the bog but no DNA. *Nature* 323: 208.

Huntington, R. 1987. *Gender and Social Behaviour in Madagascar*. Bloomington, Indiana University Press.

Huntington, R. and Metcalf, P. 1979. *Celebrations of Death: the Anthropology of Mortuary Ritual*. Cambridge, Cambridge University Press.

Hyde, H.M. 1971. *Stalin: the history of a dictator*. London, Rupert Hart-Davis.

Isbell, W.H. 1997. *Mummies and Mortuary Monuments: a post-processual prehistory of Central Andean social organization*. Austin, University of Texas Press.

Iserson, K.V. 1994. *Death to Dust: What Happens to Dead Bodies*. Tucson, Galen Press.

Janaway, R. 1996. The decay of buried human remains and their associated materials. In J. Hunter, C. Roberts and A. Martin (eds), *Studies in Crime. An Introduction to Forensic Archaeology*. London, Batsford, pp. 58–85.

Janssens, P.A. 1957. Medical views on prehistoric representations of human hands. *Medical History* 1: 318–22.

Jupp, P.C. and Walter, T. 1999. The healthy society: 1918–98. In P.C. Jupp and C. Gittings (eds), *Death in England: an illustrated history*. Manchester, Manchester University Press, pp. 256–82.

Kamp, K.A., Timmerman, N., Lind, G., Graybill, J. and Natowsky, I. 1999. Discovering childhood: using fingerprints to find children in the archaeological record. *American Antiquity* 64: 309–15.

Kantorowicz, E. 1957. *The King's Two Bodies*. Princeton, Princeton University Press.

Keenleyside, A., Song, X., Chettle, D.R. and Webber, C.E. 1996. The lead content of human bones from the 1845 Franklin expedition. *Journal of Archaeological Science* 23: 461–5.

Koch, E. 1998. *Neolithic Bog Pots from Zeeland, Møn, Lolland and Falster*. Copenhagen, Nordiske Fortidsminder B16.

Koch, E. 1999. Neolithic offerings from the wetlands of eastern Denmark. In B. Coles, J. Coles and M. Schou Jørgensen (eds), *Bog Bodies, Sacred Sites and Wetland Archaeology*. Exeter, WARP, University of Exeter, pp. 125–31.

Kowal, W.A., Beattie, O.B., Baadsgaard, H. and Krahn, P.M. 1990. Did solder kill Franklin's men? *Nature* 343: 319–20.

Kowal, W.A., Beattie, O.B., Baadsgaard, H. and Krahn, P.M. 1991. Source identification of lead found in tissues of sailors from the Franklin Arctic expedition of 1845. *Journal of Archaeological Science* 18: 193–203.

Krogman, W.M. and Iscan, M.Y. 1986. *The Human Skeleton in Forensic Medicine*. Springfield, Thomas.

Kuehn, D.D. 1999. Frozen in time: a prehistoric hunter rises from a Canadian glacier. *Discovering Archaeology* 6: 78–81.

Layton, R. 1992. *Australian Rock Art, a New Synthesis*. Cambridge, Cambridge University Press.

Leakey, M.D. 1986. The hominid footprints. In M.D. Leakey and J. Harris, *Laetoli: a Pliocene Site in Northern Tanzania*. Oxford, Oxford University Press, pp. 490–502.

Leakey, M.D. and Hay, R.L. 1979. Pliocene footprints in the Laetolil Beds at Laetoli, northern Tanzania. *Nature* 278: 317–23.

Lehner, M. 1997. *The Complete Pyramids*. London, Thames and Hudson.

Lillie, M. 1999. Cranial surgery dates back to mesolithic. *Nature* 391: 854.

Lindsay, J. 1960. *The Writing on the Wall*. London, Frederick Muller Ltd.

Litten, J. 1991. *The English Way of Death: the Common Funeral since 1450*. London, Hale.

Lobdell, J.E. and Dekin, A.A. 1984. The frozen family from the Utqiagvik site, Barrow, Alaska. *Arctic Anthropology* 21: 1–154.

Lucas, A. 1932. The use of natron in mummification. *Journal of Egyptian Archaeology* 18: 125–40.

Lucas, A. 1962. *Ancient Egyptian Materials and Industries*. Fourth edition. London, Edward Arnold.

Mack, J. 1986. *Madagascar: Island of the Ancestors*. London, British Museum.

Magilton, J.R. 1995. Lindow Man: the Celtic tradition and beyond. In R.C. Turner and R.G. Scaife (eds), *Bog Bodies: New Discoveries and New Perspectives*. London, British Museum Press, pp. 183–7.

Main, M. 1977. *Evita: the woman with the whip*. London, Corgi.

Mallory, J.P. and Mair, V.H. 2000. *The Tarim Mummies*. London, Thames and Hudson.

Mania, D. and Toepfer, V. 1973. Königsaue. Gliederung, Ökologie und Mittelpaläolithische Funde der letzen Eiszeit. *Veröffenlichungen des Landesmuseums für Vorgeschichte in Halle* 26. Berlin.

Mann, R.W., Bass, W.M. and Meadows, L. 1990. Time since death and decomposition of the human body: variables and observations in case and experimental field studies. *Journal of Forensic Sciences* 35: 103–111.

Mant, A.K. 1987. Knowledge acquired from post-war exhumations. In A. Boddington, A.N. Garland and R.C. Janaway (eds), *Death, Decay and Reconstruction*. Manchester, Manchester University Press, pp. 65–78.

Margetts, E.L. 1967. Trepanation of the skull by the medicine-man of primitive cultures, with particular reference to present-day native East African practice. In D. Brothwell and A.T. Sandison (eds), *Diseases in Antiquity*. Springfield, Charles C. Thomas, pp. 673–701.

May, J. 1996. *Dragonby: report on excavations at an Iron Age and Romano-British settlement in North Lincolnshire*. Oxford, Oxbow.

McEwan, C. and Van de Guchte, M. 1992. Ancestral time and sacred space in Inca state ritual. In R.F. Townsend (ed.), *The Ancient Americas: art from sacred landscapes*. Chicago and Munich, Art Institute of Chicago, pp. 359–71.

Medina Monzón, J.L. 1993. *Las Momias Naturales: el por qué del fenómeno de la momificación natural*. Vigo, Ediciones Cardeñoso.

Metcalf, P. and Huntington, R. 1991. *Celebrations of Death: the Anthropology of Mortuary Ritual*. Second edition. Cambridge. Cambridge University Press.

Mitford, J. 1965 and 1998. *The American Way of Death*. London, Hutchinson.

Moore, D.R. 1977. The hand stencil as symbol. In P.J. Ucko (ed.), *Form in Indigenous Art*. London, Duckworth, pp. 318–24.

Munksgaard, E. 1984. Bog bodies – a brief survey of interpretations. *Journal of Danish Archaeology* 3: 120–23.

Neely, M.E. 1982. *The Abraham Lincoln Encyclopedia*. New York, McGraw Hill.

Nelson, R.F. 1968. *We Froze the First Man*. New York, Dell.

Oakley, K.P. 1960. Ancient preserved brains. *Man* 121: 90–91.

Ó Floinn, R. 1995. Recent research into Irish bog bodies. In R.C. Turner and R.G. Scaife (eds), *Bog Bodies: New Discoveries and New Perspectives*. London, British Museum Press, pp. 137–45.

O'Sullivan, D. 1982. St Bees man: the discovery of a preserved medieval body in Cumbria. Middleberg, *Proceedings of the Paleopathology Association Fourth European Meeting*, pp. 171–7.

Pääbo, S. 1985a. Molecular cloning of ancient Egyptian mummy DNA. *Nature* 314: 644–5.

Pääbo, S. 1985b. Preservation of DNA in ancient Egyptian mummies. *Journal of Archaeological Science* 12: 411–17.

Painter, T.J. 1995. Chemical and microbiological aspects of the preservation process in Sphagnum peat. In R.C. Turner and R.G. Scaife (eds), *Bog Bodies: New Discoveries and New Perspectives*. London, British Museum Press, pp. 88–99.

Panagiotakopulu, E. and Buckland, P.C. 2001 (in press). Rameses II and the tobacco beetle. *Antiquity*.

Parker Pearson, M. 1982. Mortuary practices, society and ideology: an ethnoarchaeological study. In I. Hodder (ed.), *Symbolic and Structural Archaeology*. Cambridge, Cambridge University Press, pp. 99–113.

Parker Pearson, M. 1984. Economic and ideological change: cyclical growth in the pre-state societies of Jutland. In D. Miller and C. Tilley (eds), *Ideology, Power and Prehistory*. Cambridge, Cambridge University Press, pp. 69–92.

Parker Pearson, M. 1986. Lindow Man and the Danish connection. *Anthropology Today* 2: 15–18.

Parker Pearson, M. 2000. *The Archaeology of Death and Burial*. Stroud, Sutton.

Partridge, R.B. 1994. *Faces of the Pharaohs: royal mummies and coffins from ancient Thebes*. London, Rubicon Press.

Payne, J.A. 1965. A summer carrion study of the baby pig *Sus scrofa* Linnaeus. *Ecology* 46: 592–602.

Pizarro, P. [1571] 1978. *Relación del Descubrimiento y Conquista del Perú*. Lima, Pontificia Universidad Católica del Perú.

Polosmak, N. 1994. A mummy unearthed from the Pastures of Heaven. *National Geographic* 186 (4): 80–103.

Polson, C.J. and Marshall, T.K. 1975. *The Disposal of the Dead*. Third edition. London, English Universities Press.

Power, J.C. 1872. *Abraham Lincoln: his great funeral cortege, from Washington City to Springfield, Illinois, with a history and description of the National Lincoln Monument*. Springfield, privately printed.

Pretty, G.L. 1969. The Macleay Museum mummy from Torres Straits: a postscript to Elliot Smith and the diffusion controversy. *Man* 4: 24–43.

Pretty, G.L. and Calder, A. 1998. Mummification in Australia and Melanesia. In A. Cockburn, E. Cockburn and T.A. Reyman (eds), *Mummies, Disease and Ancient Cultures*. Second edition. Cambridge, Cambridge University Press, 289–307.

Pyatt, F.B., Beaumont, E.H., Lacy, D., Magilton, J.R. and Buckland, P.C. 1991. *Non Isatis sed Vitrum* or the colour of Lindow Man. *Oxford Journal of Archaeology* 10: 61–73.

Pyatt, F.B., and Storey, D.M. 1995. Mobilisation of elements from the bog bodies Lindow II and III, and some observations on body painting. In R.C. Turner and R.G. Scaife (eds), *Bog Bodies: New Discoveries and New Perspectives*. London, British Museum Press, pp. 62–75.

Radzinsky, E. 1996. *Stalin*. London, Hodder and Stoughton.

Reeve, J. and Adams, M. 1993. *The Spitalfields Project. Volume 1: The Archaeology: Across the Styx*. York, Council for British Archaeology.

Reeve, J. and Cox, M. 1999. Research and our recent ancestors: post-medieval burial grounds. In J. Downes and T. Pollard (eds), *The Loved Body's Corruption: archaeological contributions to the study of human mortality*. Glasgow, Cruithne Press, pp. 159–70.

Reid, H. 1999. *In Search of the Immortals: mummies, death and the afterlife*. London, Headline.

Reinhard, J. 1996. Peru's ice maidens. *National Geographic* 189: 62–81.

Reinhard, J. 1997. Mummies of Peru. *National Geographic* 191: 36–43.

Remnick, D. 1994. *Lenin's Tomb: the last days of the Soviet empire*. London, Penguin.

Reyman, T.A., Nielsen, H., Thuesen, I. *et al.* 1998. New investigative techniques. In A. Cockburn, E. Cockburn and T.A. Reyman (eds), *Mummies, Disease and Ancient Cultures*. Second edition. Cambridge, Cambridge University Press, pp. 353–94.

Robbins, L.M. 1986. Hominid footprints from Site G. In M.D. Leakey and J. Harris (eds), *Laetoli: a Pliocene Site in Northern Tanzania*. Oxford, Oxford University Press, pp. 497–502.

Roberts, G., Gonzalez, S. and Huddart, D. 1996. Intertidal Holocene footprints and their archaeological significance. *Antiquity* 70: 647–51.

Rodriguez, W.C. and Bass, W.M. 1983. Insect

activity and its relationship to decay rates of human cadavers in East Tennessee. *Journal of Forensic Sciences* 28: 423–32.

Rodriguez, W.C. and Bass, W.M. 1985. Decomposition of buried bodies and methods that may aid in their detection. *Journal of Forensic Sciences* 30: 836–52.

Rollo, F., Luciani, S., Canapa, A. and Marota, I. 2000. Analysis of bacterial DNA in skin and muscle of the Tyrolean iceman offers new insight into the mummification process. *American Journal of Physical Anthropology* III: 211–19.

Ross, A. and Robins, D. 1989. *The Life and Death of a Druid Prince*. London, Rider.

Royal, W. and Clark, E. 1960. Natural preservation of human brain, Warm Mineral Springs, Florida. *American Antiquity* 26: 285–7.

Rudenko, S.I. 1970. *Frozen Tombs of Siberia: the Pazyryk burials of Iron Age horsemen*. London, Dent.

Schmid, W.D. 1982. Survival of frogs at low temperature. *Science* 215: 697–8.

Schreiber, C. 1996. Chinchorro – the oldest mummies in the world. In P. Bahn (ed.), *Tombs, Graves and Mummies*. London, Weidenfeld and Nicolson, pp. 180–81.

Seidler, H., Bernhard, W., Teschler-Nicola, M., *et al.* 1992. Some anthropological aspects of the prehistoric Tyrolean ice man. *Science* 258: 455–7.

Sheskin, A. 1979. *Cryonics: a Sociology of Death and Bereavement*. New York, Irvington.

Sillar, W. 1992. The social life of the Andean dead. *Archaeological Review from Cambridge* 11 (1): 107–23.

Skaarup, J. 1985. *Yngre Stenalder på øerne syd for Fyn*. Rudkøbing, Langelands Museum.

Smart, N. 1998. *The World's Religions*. Second edition. Cambridge, Cambridge University Press.

Smith, C.I., Chamberlain, A.T., Riley, M.S., Cooper, A., Stringer, C.B. and Collins, M.J. 2001. Neanderthal DNA; not just old but old and cold. *Nature*: 410: 771–2.

Smith, G.S. and Zimmerman, M.R. 1975. Tattooing found on a 1600 year old frozen mummified body from St Lawrence Island, Alaska. *American Antiquity* 40: 434–7.

Snape, S. 1996. Making mummies. In P. Bahn (ed.), *Tombs, Graves and Mummies*. London, Weidenfeld and Nicolson, pp. 182–5.

Spencer, A.J. 1982. *Death in Ancient Egypt*. Harmondsworth, Penguin.

Spindler, K. 1994. *The Man in the Ice*. London, Weidenfeld and Nicholson.

Springfield, A.C., Cartmell, L.W., Aufderheide, A.C., Buikstra, J. and Ho, J. 1993. Cocaine and metabolites in the hair of ancient Peruvian coca leaf chewers. *Forensic Science International* 63: 269–75.

Stead, I.M., Bourke, J.B. and Brothwell, D. (eds) 1986. *Lindow Man: the Body in the Bog*. London, British Museum Publications.

Tapp, E. 1985. St Bees man. *Paleopathology Newsletter* 52: 10–13.

Taylor, J.H. 2001. *Death and the Afterlife in Ancient Egypt*. London, British Museum Press.

Terrill, R. 1980. *Mao: a biography*. New York, Harper and Row.

Thorpe, I.J. 1996. *The Origins of Agriculture in Europe*. London, Routledge.

Tkocz, I., Bytzer, P. and Bierring, F. 1979. Preserved brains in medieval skulls. *American Journal of Physical Anthropology* 51: 197–202.

Tumarkin, N. 1983. *Lenin Lives! The Lenin cult in Soviet Russia*. Cambridge MA, Harvard University Press.

Turner, R.C. 1995. Recent research into British bog bodies. In R.C. Turner and R.G. Scaife (eds), *Bog Bodies: New Discoveries and New Perspectives*. London, British Museum Press, pp. 108–22.

Turner, R.C. 1999. Dating the Lindow Moss and other British bog bodies. In B. Coles, J. Coles and M. Schou Jørgensen (eds), *Bog Bodies, Sacred Sites and Wetland Archaeology*. Exeter, WARP, University of Exeter, pp. 227–34.

van der Sanden, W.A.B. 1995. Bog bodies on the continent: the developments since 1965, with special reference to the Netherlands. In R.C. Turner and R.G. Scaife (eds), *Bog Bodies: New Discoveries and New Perspectives*. London, British Museum Press. 146–65.

van der Sanden, W.A.B. 1996. *Through Nature to Eternity. The Bog Bodies of Northwest Europe*. Amsterdam, Batavian Lion International.

Vreeland, J.M. 1978. Prehistoric Andean mortuary practice: preliminary report from Peru. *Current Anthropology* 19: 212–14.

Vreeland, J.M. 1998. Mummies of Peru. In A. Cockburn, E. Cockburn and T.A. Reyman (eds), *Mummies, Disease and Ancient Cultures*. Second edition. Cambridge, Cambridge University Press, pp. 154–89.

Wallis Budge, E.A. [1893] 1987. *The Mummy: a handbook of Egyptian funerary archaeology*. London, Kegan Paul.

Walsh, G.L. 1979. Mutilated hands or signal stencils? *Australian Archaeology* 9: 33–41.

Waugh, E. 1948. *The Loved One: an Anglo-American Tragedy*. Harmondsworth, Penguin.

Wilmut, I., Schnieke, A.E., McWhir, J., Kind, A.J. and Campbell, K.H.S. 1997. Viable offspring derived from fetal and adult mammalian cells. *Nature* 385: 810–13.

Zimmerman, M.R. and Aufderheide, A.C. 1984. The frozen family of Utqiagvik: the autopsy findings. *Arctic Anthropology* 21: 53–64.

Zimmerman, M.R. and Smith, G.S. 1975. A probable case of accidental inhumation 1600 years ago. *Bulletin of the New York Academy of Medicine* 51: 828–37.

Photographic acknowledgements

COLOUR ILLUSTRATIONS

I. Professor Dr Gunther von Hagens, Institute for Plastination, Heidelberg, Germany
II. Novosti, London
III. Silkeborg Museum, Denmark
IV. Schleswig-Holsteinische Landesmuseen Schloss Gottorf
V. Chris Sharp/South American Pictures
VI. Tony Morrison/South American Pictures
VII. The British Museum
VIII. Sygma
IX. Owen Beattie
X. Paul Hanny/Frank Spooner Pictures
XI. Johan Reinhard
XII. Maureen Carroll
XIII. Jean Vertut
XIV. Linares Family, Mexico City/The British Museum

BLACK-AND-WHITE
ILLUSTRATIONS

1. The British Museum
2. © Steve Hopkin
3. Andrew Chamberlain
4. Antonio Ascenzi
5. Durham Cathedral
6. By permission of the British Library (Add.49598.f.90)
7. Dean and Chapter of Westminster
8. Musée de la Tapisserie, Bayeux, France/The Bridgeman Art Library
9. By courtesy of the National Portrait Gallery
10. From Guamán Poma [1613] 1956
11. Illinois State Historical Library

12. Novosti, London
13. Novosti, London
14. Novosti, London
15. Associated Press
16. © Roberto Adaniya/Andes Press Agency
17. Jorn St Jerneklar/Impact
18. Professor Paul Buckland, University of Sheffield
19. Andrew Chamberlain
20. Forhistorisk Museum, Moesgård, Denmark
21. Silkeborg Museum
22. The National Museum of Denmark
23. The National Museum of Denmark
24. The National Museum of Denmark
25. The National Museum of Denmark
26. Forhistorisk Museum, Moesgård, Denmark
27. The National Museum of Denmark
28. Silkeborg Museum
29. Forhistorisk Museum, Moesgård, Denmark
30. The British Museum
31. The British Museum
32. Don Brothwell and Jim Bourke
33. Andrew Chamberlain
34. Drents Museum (J. Bosma)
35. Michael Parker Pearson
36. National Museum of Wales
37. British Library, Aylett Sammes Britannia Antiqua 1676 (C83K2)
38. Royal British Columbia Museum
39. Deirdre O'Sullivan
40. © Chris Sharp/South American Pictures
41. By permission of the British Academy
42. Ashmolean Museum, Oxford

43. The British Museum
44. Hierakonpolis Expedition Archives; photographer R. Friedman
45. The British Museum
46. Cairo Museum
47. The British Museum
48. The Griffith Institute, Ashmolean Museum, Oxford
49. Museo Arqueológico de Tenerife (OAMC)
50. © Tony Morrison/South American Pictures
51. Pat Remler
52. Christine Osborne Pictures
53. Victor H. Mair
54. Mary Evans Picture Library
55. Owen Beattie
56. Werner Forman Archive/The Greenland Museum
57. Werner Forman Archive/The Greenland Museum
58. The British Museum
59. Paul Hanny/Frank Spooner Pictures
60. Barbara Ottaway
61. Johan Reinhard
62. Novosti, London
63. Alex Norman, after Rudenko 1970
64. A.A. Gavrilovoi
65. Sergei I. Rudenko
66. Novosti, London
67. Novosti, London
68. The British Museum
69. The British Museum
70. Michael Parker Pearson
71. Maureen Carroll
72. Maureen Carroll
73. Ronald Sheridan/Ancient Art and Architecture Collection Ltd
74. David Keith Jones, Images of Africa Photobank
75. David Keith Jones, Images of

Africa Photobank
76. Andrew Chamberlain
77. Andrew Chamberlain
78. Jacques Collina-Girard
79. Jean Vertut
80. Ulmer Museum/Thomas Stephan
81. Jean Vertut
82. University College London/Bentham Collection
83. Andrew Chamberlain
84. Greenhaven Woodland Burial Ground Ltd
85. PA News
86. Dr José Luis Medina Monzón and Ediciones Cardenoso
87. The British Museum
88. Olivia Forge; supplied by Pete Stone
89. J. Nolan

Index